FIRE IN THE FOREST

How destructive or beneficial are forest fires to wildlife? Should we be trying to reduce or increase the amount of fire in forests? How are forest fires controlled, and why does this sometimes fail? What effect will climate change have? These and many other questions are answered in this richly illustrated book, written in non-technical language. The journey starts in the long geological history of fire, leading up to our present love--hate relationship with it. Exploring the physics of how a single flame burns, the journey continues through how whole forests burn and the anatomy of firestorms. The positive and negative ecological effects of fires are explored, from plants and wildlife to whole landscapes. The journey ends with how fires are controlled, and a look to the future. This book will be of interest to ecologists, biogeographers and anyone with an interest in forest fires and the role they play.

Peter A. Thomas is Senior Lecturer in Environmental Science at Keele University, UK and has been a Visiting Professor at the University of Alberta, Canada and a Bullard Fellow at Harvard University, USA. His teaching encompasses a wide range of tree- and woodland-related topics including fire behaviour and fire ecology, and he recently received an Excellence in Teaching Award from Keele University. He is the author of *Trees: Their Natural History* and *Ecology of Woodlands and Forests*, both published by Cambridge University Press.

Robert S. (Rob) McAlpine works with the Ontario Ministry of Natural Resources, Aviation and Forest Fire Management Branch, where he leads the Science and Technology group. His current research interests and activities span fire behaviour, fire economics, equipment development and organisational effectiveness. He has worked in fire management and research for over 30 years, from a frontline firefighter to a fire scientist.

Fire in the Forest

Peter A. Thomas and Rob McAlpine

with contributions from
Kelvin Hirsch and Peter Hobson

CAMBRIDGE UNIVERSITY PRESS
Cambridge, New York, Melbourne, Madrid, Cape Town, Singapore,
São Paulo, Delhi, Dubai, Tokyo, Mexico City

Cambridge University Press
The Edinburgh Building, Cambridge CB2 8RU, UK

Published in the United States of America by
Cambridge University Press, New York

www.cambridge.org
Information on this title: www.cambridge.org/9780521822299

© Cambridge University Press 2010

First published 2010

Printed in the United Kingdom at the University Press, Cambridge

A catalogue record for this publication is available from the British Library

Library of Congress Cataloging-in-Publication Data

Thomas, Peter A.
    Fire in the forest / Peter A. Thomas and Rob McAlpine ; with contributions from
Kelvin Hirsch & Peter Hobson.
        p.   cm.
    ISBN 978-0-521-82229-9 (Hardback)
    1. Forest fires.   I. McAlpine, Rob.   II. Title.
    SD421.T49 2010
    634.9′618–dc22   2010026385

ISBN 978-0-521-82229-9 Hardback

To our families who have stood with us during this project:
Caroline, Benjamin and Graeme (RSM); Judy, Matthew and Daniel (PAT).

Contents

Preface

Year after year pictures in the media show towering flames threatening people's homes, livelihoods, and their very lives in places as diverse as North America, Europe and Australia -- why does this happen? Conflicting stories continually appear over whether fire is rapidly destroying the animals, habitats and plants we treasure in our forests, or whether fire is their salvation, the key to diversity and ecosystem rebirth -- where does the truth lie? With global warming predictions, do we face more and larger fires or will technology be able to tame this potentially savage enemy? This book delves into these and other questions, providing a factual account and perspective of how fire burns in the forest, what it does and how it might be controlled.

Where the published work of others is used, or where good sources of extra information are recommended, the authors and the date of the publication are given so that the source can be found in the references at the end of the book. This inevitably has resulted in a compromise; we've tried to keep this to a minimum to help the text flow but give sufficient references to help the reader who wishes to find out more. Our apologies if we fail you at any point.

We are indebted to the many people who provided information and insights and who read part or all of this book. In particular we are grateful to Dave Bowman, Yeonsook Choung, Helene Cleveland, Malcolm Gill, Jim Gould, Richard Hobbs, Scott Keelan, John Packham, Marc-André Parisien, Steve Pyne, Tim Sheldan, Brian Stocks, Jan Volney, Mike Weber, Tim Williamson and Mike Wotton. People were also remarkably kind in sharing pictures and photographs with us; these are gratefully acknowledged in the figure legends. All other photographs were taken by PAT. We are very grateful to Andy Lawrence who did his usual superb job in drawing many of the colour figures. Despite the help of others, if errors remain they are our fault. Please do let us know where you do find errors or you disagree with the stance taken.

PAT is happy to record that some of the background for this book was researched while a Bullard Fellow at Harvard University, Massachusetts.

Contributors

MAIN AUTHORS

Peter A. Thomas
School of Life Sciences
Keele University
Keele
Staffordshire ST5 5BG, UK

Rob McAlpine
Fire Science and Technology
Aviation and Forest Fire Management
Ontario Ministry of Natural Resources
70 Foster Drive, Suite 400
Sault Ste. Marie, Ontario,
Canada P6A 6V5

CONTRIBUTING AUTHORS

Peter Hobson
School of Sustainable Environments
Writtle College
Chelmsford
Essex CM1 3RR, UK

Kelvin Hirsch
Canadian Forest Service
Northern Forestry Centre
5320–122nd Street
Edmonton, Alberta,
Canada T6H 3S5

1 | In the beginning

Welcome to *Fire in the Forest*. The ancient Greeks considered fire as one of the classical elemental forces along with water, earth and air; and indeed fire has helped shape the world around us to such a great extent that it would be hard to argue the point. Without a doubt, there is nothing on this planet that cannot be traced back (many times over) to some kind of fiery origin, whether that be the Big Bang at the origin of the universe, the atomic fires of some long dead pre-supernova star, the liquid burning rock just below the crust of the earth or the vegetation sitting fraily on the surface of the planet. While all of these forging fires are fascinating, a tome that spans the fullness of fire would be many volumes thick. The focus of this book is to explore the various facets of fires in forests: from how a single flame works, to what determines whether a forest will burn, to why huge forest fires occur and how they can be tackled. In this we will be dealing primarily with wildland fires – fires in the natural or semi-natural forest rather than those in plantations or in urban areas. These wildland fires have shaped the planet's biotic structure so we also focus on how plants and animals cope with fire, our own interaction with fire and ultimately our overall relationship with the whole planet. We hope that this book will be of use to anyone with an interest in how forest fires work and how they affect the forest. Our aim has been to make this sometimes complex world open and approachable to anyone with or without even a hazy knowledge of science.

When the first plants began to colonise the land of the early earth they adapted to the variously forming environmental niches. Into this inhospitable place, plants took hold and grew, but fire was there too. During rainless spells, when the land and plants dried out a little, lightning and even volcanoes would ignite fires. These fires would spread out, consuming the plants and releasing the nutrients back to the environment. Following the fires, more plants arose, utilising the nutrients that were now more freely available. This cycle was repeated almost endlessly, each successive cycle consuming another generation of plant life, and slowly changing the environment. In some climate zones, seasonal moisture would produce lush growth only to have it dry out at another time of the year resulting in frequent, sometimes annual burning, while other locations, due to a lack of fuel accumulation, or of near-constant fire-prohibiting moisture, or an infrequent ignition source, would see fire much less often. These climatic effects began to be drivers of evolution for plants, with species beginning to adapt to different degrees of burning and exploit the elemental force of fire as they would the other elements.

From the beginning our (human) relationship with fire has been two-sided. We exploited the use of fire for hunting and warmth, and our safety was threatened by fire. As civilisation evolved, our relationship with fire included cooking, forging of metals and further. However, this evolution eventually began to weaken our understanding of fire as a natural process. As civilisation has grown, fire has been increasingly containerised, first into the hearth then into internal combustion engines and power-generating plants, so we lost our experience and appreciation of the benefits of open fires while keeping our fear. Thus, to many, fire in the landscape became a wholly negative thing (more on that later!).

If we fast forward to today, we find fire is still an essential natural process in many ecosystems across the planet. Plants have evolved since they first colonised the land, not only into more complex organisms adapted to the weather and climate of various regions, but also adapted to the frequency and intensity of fires. Indeed many plants and animals have evolved to influence the fire cycle and to use fire to their own ends, thereby out-competing other organisms not so well adapted to fire, or adapted to a different fire regime. The range and ingenuity of these adaptations is as varied as life itself. Certainly we know that the world would look very different without fire. If we could completely remove all fire, grasslands would diminish greatly in size, and forest cover would increase from 30% to an estimated 56% of the vegetated land surface (Bond *et al.* 2004, Bowman 2005).

In this single volume we could not begin to address the wide range of fire effects and cover the breadth of fire management across the globe. Instead we have chosen some key fire-prone areas around the planet – the global fire powers of Canada, USA and Australia – and focused on them, with occasional diversions to neighbouring places for illustrative purposes. The boreal forest, dominated by pines and spruces, is a large fire-prone ecosystem, encircling the northern hemisphere. Over a third of the boreal forest lies within Canada and here lies one of our areas of focus. The USA contains a good deal of even more flammable forest vegetation which has its own story to tell. Australia is an island continent that perhaps lives with the most frequent and intense fire regime on the planet. From the arid grassland to the eucalypt forests, fire is an integral part of the land and the people.

The nature of fire

Within the ancient Greeks' four elemental forces, all except fire are matter based – earth, air and water are things that today are understood to be composed of matter. Only fire is different, and rather than an 'object' so to speak, it is a process. Certainly the other elements can be involved in processes (water for example can be an agent of change in erosion), but it is only fire that is really intangible and ephemeral. At the same time, fire is the only one of these elemental forces that we can unleash and to a certain extent control as a major ecological force. We can light fires but cannot start a volcanic eruption, hurricane or widespread flood. These facts demand a better understanding of what fire is. In chemical terms, fire is an oxidising agent – it combines one lump of matter with oxygen to produce another form of matter. Oxidation is often quite slow (as when oxidised iron turns to rust, for example) but in other cases, such as fire, it is a more rapid process. Some forms of matter when oxidised (combined with oxygen) react and release heat. In some cases (phosphorus for example) no heat is required to start this reaction, but in most cases (including fire) heat is required to initiate the reaction. The extra heat then generated by oxidation is sufficient to sustain the process until all the matter is transformed (or in the case of wood – fuel consumed).

To lay a broad background, fire is in essence the opposite of photosynthesis. Plants grow using photosynthesis to capture the energy of the sun and convert carbon dioxide and water into glucose (for plant structures) and release oxygen and water back into the atmosphere.

A simplified equation for photosynthesis is:

$$6CO_2 + 12\,H_2O + \text{sunlight yields } C_6H_{12}O_6 + 6\,O_2 + 6\,H_2O.$$

Or the same equation in recipe format:

> 6 parts carbon dioxide (CO_2) +
> 12 parts water (H_2O) +
> sunlight

Yields:

> One part glucose ($C_6H_{12}O_6$) +
> 6 parts oxygen (O_2) +
> 6 parts water (H_2O).

Of course, plants are composed of more than glucose, but glucose is the major building block of most structures (e.g. cellulose, wood) and for our purposes the essential element of concern. Additionally, there are a host of micronutrients and minerals required for the health of the plant (supplied by the soil). Where photosynthesis converts water and carbon dioxide into glucose, fire reverses this reaction, oxidising the glucose back into the original constituents and releasing the stored solar energy as heat and light. The difference between photosynthesis and fire (other than one being the reverse of the other) is the rate at which the two occur. A large tree (and by extension forest) collects solar energy over decades, centuries or millennia, while the stored energy can be released in minutes during a fire; consequently the energy release rates can be enormous. On a large, high-intensity wildfire, it is said that the energy equivalent of one Hiroshima-sized atomic bomb can be released every 20 minutes. It is worth remembering that the energy released during fire was captured from the sun and stored by the trees and vegetation; nature's first energy storage battery.

Looking back at the Greek philosophers, they seemed again to get things right – earth, air and water combine to form the trees and other vegetation with energy from the sun (at its heart an atomic fire). And when the process is reversed, the energy initially forged from the sun, and translated into form by trees and vegetation, is released by fire to consume the storage container and yield heat and light.

Just how widespread are forest fires?

A common question asked about fire is how much of the world's land surface is affected each year. Satellite imagery and ground-based estimates suggest that between 2000 and 2004 (a fairly normal period) the area burnt varied between 2.97 and 3.74 million km^2 each year, most of this in forests (Giglio *et al.* 2006, FAO 2007). This area, approximately the size of India, burnt each year you can see isolated in dark colours in Fig. 1.1. Or if you prefer, this is equivalent to just under a third of Canada, a little less of the USA, 40% of Australia or 12 times the area of the UK.

The next question usually asked is which parts of the world burn most often. Today across the planet, biomes (major vegetation types) have fallen into balance with the local climatic conditions, the soil richness and the preponderance of fire. There are few places that are immune to fire (Antarctica being one) but of a higher importance perhaps is the frequency that fires return to an area. This fire return interval within an area impacts the vegetation most profoundly.

Fig. 1.1 begins to describe a global picture of fire frequency. Darker areas of the map have a greater percentage of the 1° × 1° (latitude by longitude) cells burnt by fire annually, indicating a higher fire frequency. For fires of natural origin (that is, not started by humans – see Chapter 3) we see the darkest areas in the world's grasslands and savannahs. Unfortunately, Fig. 1.1 may be distorted from a 'natural'

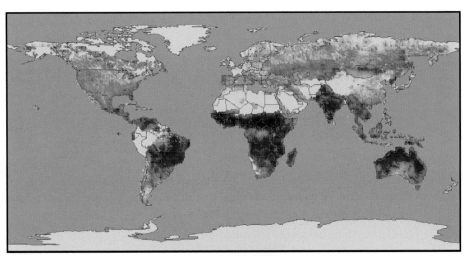

0% burnt [gradient bar] > 60% burnt

Figure 1.1
A 1° × 1° global map of average annual area burnt (% of cell burnt) for 1960–2000. © International Association of Wildland Fire 2009. Reproduced with permission from the *International Journal of Wildland Fire* **18**(5): 483–507 (Flannigan, M.D., Krachuck, M.A., de Groot, W.J., Wotton, M & Gowman, L.M.) DOI 10.1071/WF08187. Published by CSIRO Publishing, Melbourne, Australia.

fire cycle by tropical rainforest burning activities in South America and South-East Asia. Although the tropics are often too wet to burn, as described in Chapter 4, people clearing the forest, aided by El Niño droughts, let in fire. These can be huge fires: the Great Fire of Borneo between September 1982 and July 1983 burnt between 35 and 37 thousand km^2 – a hundredth of the normal global area burnt per year in just one fire (Johnson 1984).

However, a striking fact clearly visible from Fig. 1.1 is that generally where there is vegetation, there is fire (with few exceptions). The vast majority of these fires are not monitored or documented due to remoteness and so go largely unnoticed by the wider world. Some grassland areas burn almost every year (consequently keeping them grasslands!). This link between grasslands and high fire frequency begins to show us the linkage between the two – repeated fires on a near-annual basis do not allow longer-lived organisms to survive – they are repeatedly burnt back and not allowed to get a start. So the grassland ecosystem, home to many species of both flora and fauna, is dependent on frequent fire to maintain its character. Another example, although very different ecologically, is the eucalypt forests in many parts of Australia (Fig. 1.2). And as we see later in Chapter 6, most ecosystems are dependent on some level of fire frequency.

However, this understanding of the necessary role of fire in the ecosystem is only a relatively recent revelation for modern civilisation. This raises the perennial question of whether forest fires are indeed friend or foe. It was not until the 1960s that the essential role of fire was first suggested, and acceptance into the main-stream has taken decades. Even in the 1970s fire was mostly thought to be a bad thing: a Canadian Forestry Service publication called *Forest Enemies* (Anon 1973) included a section on fire. At the same time, the Canadian company Bombardier, who build water bombers, produced an educational video that described their machines as fighting the number one enemy of the forest. But our perspective has changed and our depth of understanding of fire has increased so that we now see some fires as useful and others as harmful. So the same dichotomy of the nature of fire exists for us; fire is useful (technologically) and an essential part of the ecosystems, but unmanaged/uncontrolled it can threaten lives, structures, infra-structure (and the dependent economies), and our way of life. This produces a problem; as Putz (2003) points out:

> In the glare of the conflagrations that consume forests and kill firefighters in
> western and northern North America every fire season, special care is
> required when trying to present fire in a positive light.

We therefore try and tread carefully in this book, aiming to give you the facts on how fires work in the landscape and why and how we try to control them, and to leave you to decide how you feel about the pros and cons of different fires.

Acceptance of fire has been hampered by the long-time ingrained desire and belief that nature can be tamed and dominated. This attitude not only drove fire

Figure 1.2
Recently burnt forest in Western Australia with greening vegetation immediately post-fire. Photograph by Rob McAlpine.

and land-management policies, but facilitated attitudes that people could build homes in 'natural' environments and expect their homes to be protected from fire. Governments have met this challenge and built complex firefighting organisations. Society now depends on these firefighters to protect people, homes, communities and the infrastructure that the economy depends upon (pipelines, hydro lines, telecommunication towers, railroads, crops, timber for homes etc.).

How firefighters meet the daily challenge that nature has set out, is a story unto itself. Complex organisations utilising state-of-the-art technology, adapting to various situations, meet the challenge of fire to protect what society demands it protects. However, as in all cases when humans contend with nature, there are wins and losses. New technologies improve capabilities, at an ever-increasing price tag, but on some days, the elemental force of fire conquers all.

Fire then is a conundrum – a basic elemental force of nature that cannot be held back, with ecosystems evolved the world over that require rebirth or rejuvenation

through fire; and humans who need to protect their lives and property from the destructive force of fire. Meeting that balance will challenge us all.

The Greeks recognised the fire shown in Fig. 1.3 as an element of nature. In this book we will look at wildland fire in several key locations around the world to understand how forest fires burn (Chapters 3 and 4), how ecosystems have integrated fire into their function (Chapters 5 and 6) and how we humans meet the challenge of fire (Chapter 7). From there we may start, from looking at the modern world, to understand why fire – a process and not a physical element – became part of the Ancients' elements. And finally we look to the future (Chapter 8) and the continuing and increasingly complex challenge of finding the balance for fire in the forest.

Figure 1.3
A fire burning in the boreal forest of Northern Ontario. Ontario Ministry of Natural Resources, copyright 2006 Queens Printer Ontario.

2 | Historical review

The earliest beginnings of fire in geological time

Fires could not blaze until there was enough oxygen in the atmosphere and enough vegetation to burn. Oxygen began to appear in our atmosphere around 2 billion years ago (in the Archaean subdivision of the Precambrian Era) and reached its present-day level (21% of the atmosphere) around 600 million years ago. Life was still in the seas at this point – in fact it was only when oxygen reached these high levels that ozone could be formed high in the atmosphere which screened out harmful ultraviolet radiation and allowed complex life to invade the shallows and finally land. Oxygen levels since then have fluctuated between 15–35% of the atmosphere (Fig. 2.1). Coincidentally, fire needs a minimum of 13–15% oxygen to burn and above 35% spontaneous combustion is likely.

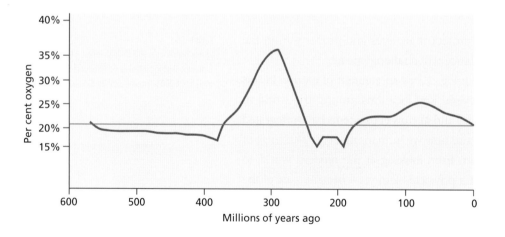

Figure 2.1
Changes in atmospheric oxygen level over the past 550 million years (the Phanerozoic Eon). The horizontal line shows the current concentration of 21%. Data from Dudley (1998) and Berner (1999).

How do we know when fires first started to appear? Fortunately, rock and coal layers contain a charcoal-like substance called fusain; small black fragments that often still contain the original cellular structure (Fig. 2.2). There has been a long debate about whether fusain is indeed a result of fire or whether it is from the much longer and slower oxidation process of weathering – see Box 2.1. But the consensus

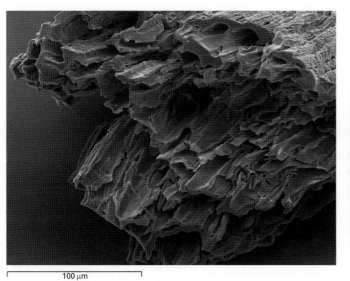

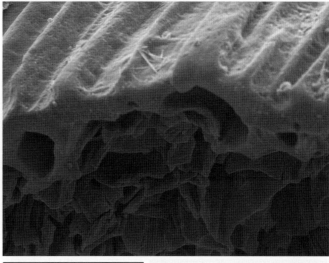

100 μm

30 μm

Box 2.1. Is all black material from burning?

Blackened plant material may immediately makes us think of burning – as in the picture on the left of Fig. 2.3. But if plant material, such as the wood of the roots and old branches on the bristlecone pine on the right, is exposed for many years to the elements it weathers to the same chemical composition and approximate appearance as charcoal. In effect, slow oxidation in weathering and fast oxidation in fire create end products that superficially look similar. Weathering, however, normally produces a thinner more superficial layer of 'char' simply because the material is likely to finally rot or break apart before it has long enough to develop deep weathering. Moreover, oxidation by fire tends to lead to melting and blistering of the waxy cuticle over the plant surface (the epidermis) which can be seen even in fossil fragments, and most importantly experiments have shown that black fragments with a high reflectance are indicative of fire temperatures of at least 400 °C and generally the greater the reflectance, the greater the fire temperature.

Figure 2.2
Charred plant fragments extracted from siltstone rocks found at Ludlow, England. These are evidence of the earliest wildfire so far documented that dates from the late Silurian (419 million years ago). The plant involved is a rhyniophyte, probably *Hollandophyton colliculum*, one of the first land plants just a few tens of centimetres tall, with leafless branched green stems and spore-producing cones. Photographs courtesy of Ian Glasspool, Dianne Edwards and Lindsey Axe.

Figure 2.3
On the left, logs burnt in a fire in New Brunswick, Canada showing extensive charring. On the right, a bristlecone pine (*Pinus aristata*) in the White Mountains, California that is probably over 4000 years old and which shows slow weathering of the exposed roots and old branches over millennia that superficially looks as though they have been burnt.

now is that the majority of this fusain does indeed result from past fires to provide clues as to the origins of fire.

Fusain deposits suggest that fires have been burning since the Silurian, 419 million years ago (see Glasspool *et al.* 2004 for further information). These fires, started by abundant lightning, would have burnt through the first land plants that appeared around 430 million years ago (early Silurian). The immense coal-producing swamps of the Carboniferous (360–290 million years ago) provided abundant material to burn with their lush forests of giant ferns, horsetails and clubmosses (Fig. 2.4). Moreover, oxygen levels had risen to a peak at this time – up to around 35% of the atmosphere (Fig. 2.1) – helping fires burn more vigorously. But was there more fire then? Well, charcoal deposits became more common through the early to late Carboniferous, including in tropical wetland peats, but there is less than expected, probably due to a wet climate putting a damper on things. However, it is tempting to think that the first signs of adaptations in plants that allowed them to cope with fire began to appear during this period (although they may have evolved initially for other reasons). For example, the living tissue buried deep in the bark of the clubmoss-trees likely made them very fire tolerant.

As new types of plants evolved they would have met with a baptism of fire. Flammable conifers became abundant in the Triassic (around 245–208 million years ago) and would have met fire (See Uhl & Montenari, 2010). Oxygen levels dropped to 15% about 200 million years ago (the Jurassic) to rise again to around current levels as the broad-leaved hardwoods appeared in the early Cretaceous (120 million years ago). Fire must have played an important role in their evolution.

It is often suggested that the extinction of the dinosaurs at the Cretaceous–Tertiary (K–T) boundary 65 million years ago was due to the meteorite impact that created the 180-km diameter Chicxulub crater on the Yucatán Peninsula in Mexico, accompanied by huge 'global wildfires', the like of which had not occurred before or since. The evidence for the impact is overwhelming but the evidence for the fires is not compelling; the soot and charcoal in the rocks of this time could have been produced by fires that burnt gradually through lots of dry, dead vegetation killed by the sudden change in climate (Belcher *et al.* 2005). Evidence for this comes from microscopic examination which shows that more than 50% of charcoal fragments from the K–T boundary show signs of rotting, compared with less than 5% observed in modern fires. Moreover, studies in North America north of the Chicxulub crater found between four and eight times *less* charcoal in K–T rocks than in rocks above and below the boundary, suggesting that the huge meteor impact resulted in relatively few fires. Also, the structure of the soot associated with the charcoal from the K–T boundary appears to come primarily from the vaporisation of hydrocarbon-rich rocks rather than burning vegetation. Nichols & Johnson (2008) give an excellent account of the vegetation either side of the boundary.

Figure 2.4
What a clubmoss forest of the Carboniferous Period may have looked like. The cone-bearing trees in the middle distance and the fallen one in the middle are *Lepidodendron* (an early clubmoss). The trunk on the right is a tree fern and some of its foliage can just be seen at the top. The dead tree on the left is an early ancestor of the conifers, and, more centrally, an unbranched species of *Sigillaria*, another clubmoss. Grasses and flowering plants had not yet evolved but mosses, ferns and small horsetails (seen behind the tree fern) were abundant. The animal in the foreground is a labyrinthodont amphibian that spent most of its life in the water but which could walk on land. On the fallen trunk is a member of the extinct 'griffenflies' which included the largest insects ever to live, and which gave rise to modern dragonflies. These huge insects and abundance of fire are both attributable to the high levels of oxygen in the atmosphere (see Fig. 2.1). Drawn by Peter R. Hobson. From Thomas & Packham (2007).

Tertiary and Quaternary – the last 65 million years

After the dinosaurs became extinct, fire continued to wax and wane as climates were more or less favourable. As our current forests – with trees we would recognise – have evolved over the last 65 million years, fire has been an integral part of that development and the world's fire-prone vegetation acquired much of its modern character. But fire has not been a passive follower of climate and vegetation; it also played a part in shaping the vegetation. This is illustrated nicely by changes during the Miocene (5–24 million years ago) when grasslands appeared around the world. Evolutionary scientists have long asked why grasslands appeared so comparatively late. One of the most likely answers is a change in world climate towards increasing aridity, especially at higher latitudes (north and south), which was needed to give grasses the competitive edge. But it was probably not just dryness; late Miocene pollen records from Africa, for example, are associated with charred grass leaf surfaces. The most likely scenario is that the drier climate led to an increase in fire frequency, which favoured those plant species, including grasses, which could cope with a dry climate, frequent fire and grazing animals.

Fire undoubtedly would have helped accelerate the change in vegetation and would have selectively encouraged those species that could cope with fire, producing a vegetation type that would otherwise not have existed. Fire was thus both an accelerant (speeding up the response to climatic change) and a selective agent, weeding out those species that could not cope.

During the unimaginable numbers of years of these long geological periods it is also important to remember that the continents have moved around the globe and so may have a fire history in geological time that is quite different from the current day. Thus fusain may occur in what can appear to be unlikely places. For example, rocks 140 million years old (early Cretaceous) in southern England show good deposits of fusain in what was then a delta of conifer woodland with a fern and clubmoss understorey. Now, southern England has little natural fire since the wet climate grows lush greenery that burns like wet asbestos except during exceptional droughts (such as the spring of 2003). Similarly, tropical areas which are now usually too wet for extensive fires – like rainforest – can have an extensive history of fire. Fire was abundant in the Amazon Basin during several dry periods over the last 2 million years, the latest being 4000 to 6000 years ago (see Sanford *et al.* 1985). However, for much of the Holocene (the last 10 000 years) this area has been in a wet phase with much less frequent fire.

Holocene – the last 10 000 years

More is known about fire during this Holocene period following the last Ice Age because most lake and bog deposits, with their trapped record of pollen and charcoal, were formed during this time. Even so, interpretation of this trapped

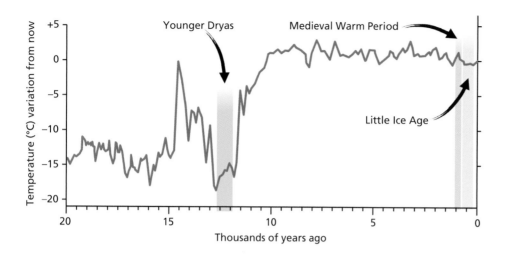

Figure 2.5
Variation in temperature over the last 20 000 years. The data are interpreted from chemical analyses of the ice and trapped gas bubbles taken along a 3000-m deep ice core from the Greenland ice cap. Temperatures at the end of the last Ice Age fluctuated widely, reaching almost current levels 14 000 years ago (the Allerød Period) before plunging into the final very cold period of the Younger Dryas. Around 11 500 years ago, the Ice Age came to an end and temperatures rose and steadied with small fluctuations to the present day. These small fluctuations are, however, significant since fire was more common during the Medieval Warm Period (roughly AD 750–1250) and much less so during the Little Ice Age (AD 1450 to c. 1850) despite the annual differences in temperature being just a few degrees. Based on data from Alley (2000, 2004).

record is not always easy. Charcoal, for example, caught in a bog may have been transported from fires considerable distances away by wind or on water (charred conifer needles float for many more days than uncharred ones). Certainly, charcoal particles from the big 1988 fires in Yellowstone were found 1600 km (1000 miles) away in Minnesota. On the positive side, palaeoecologists have a variety of methods and techniques for teasing the true picture from these deposits, and the overall picture of Holocene fire is continually improving.

As you would expect, warmer climatic periods generally led to more fire. Thus we know that there was more fire during the warmer climate of the Medieval Warm Period and less during the global period of the generally cooler and moister climate of the Little Ice Age (Fig. 2.5). Within this large overall picture, however, there were many small variations. As suggested above, the Little Ice Age in Sweden appears to have greatly reduced fire frequency above 1000 m, allowing fire-intolerant spruce to invade forests. At lower elevations in east Sweden, however, the climate was slightly warmer and drier, and fires carried on burning (Bradshaw & Zackrisson 1990).

The vegetation and animals themselves can also inject perversities into the complex relationship between fire and climate. For example, in Elk Island National Park in western Canada, the Medieval Warm Period resulted in flammable shrub birch (*Betula glandulosa*) being replaced by the less fire-prone aspen (*Populus tremuloides*), due to a lowering of the water table (see Campbell & Campbell 2000). This led to a decline in fire frequency at odds with what the climate would suggest. In the same area in more recent historical times, the amount of aspen has risen and grassland has declined (Fig. 2.6). This could be due to a reduction in fire frequency driven by climate but is most probably due to the decline in grazing animals, especially the huge historical decline in buffalo, which would previously have prevented trees from growing (Frans Vera in his book *Grazing Ecology and Forest History* (2000) looks specifically at the importance of grazing animals and their interaction with fire).

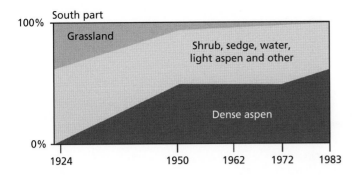

Figure 2.6
Changes in vegetation in the south part of Elk Island National Park, Alberta, Canada. The area covered by aspen has increased over recent decades. While this could be due to less fire, it is most likely due to a decline in grazing animals. Modified from Campbell & Campbell (2000), with permission from Elsevier.

The intervention of humans

The first use of fire by 'humans' probably dates back to the Australopithecines, the ancestors of modern humans, that lived at least 1.5 million years ago in the African Rift Valley. As their descendants, *Homo erectus*, spread into Europe and Asia about a million years ago and then further afield, it is clear that they took fire with them wherever they went, from jungles to savannahs. Certainly, many sites around Europe occupied by modern humans (*Homo sapiens*) 750 000 to 400 000 years ago show clear remains of fire on animal bones and fire-hardened implements. Fire was not, of course, just restricted to Europe; many people worldwide used fire, from the Aboriginals of Australia, the Maoris in New Zealand, the Melanesians in Fiji, through to the indigenous peoples of the Old World (India, Asia, Africa, Russia) and New World (Gill 1981, Kershaw *et al*. 2002, Gott 2005).

Uses of fire

The use of fire was one of the key features distinguishing early humans from the rest of the primates. We can imagine that fire was initially 'captured' from natural sources such as lightning fires and used for heat and protection from wild animals and perhaps each other. Nevertheless, it is apparent that early humans rapidly learnt to use fire as a very effective tool and indeed their life was centred around fire. Fire uses can be loosely divided into domestic and landscape.

Domestic use included heating, cooking (including preserving by drying and smoking), repelling mosquitoes, driving snakes from camping grounds, hardening wooden tools, making pottery and metal implements, protection at night and social uses (such as making the evenings available for story telling), spiritual ceremonies (see Box 2.2) and mortuary rituals.

Outside of the domestic camp, fire has been extensively used by indigenous peoples for hunting, by clearing away vegetation to make game more visible, for driving game and by encouraging young grazing to attract animals. Fire was also

Box 2.2. Spiritual and magical uses of fire

Both the Aztecs and the Incas of Peru worshipped gods of fire. The Aztecs of Mexico used huge ceremonial fires either kept burning permanently in the temples or just during major religious festivals. Much later, the Greek goddess of the hearth, Hestia, was one of the 12 Olympian deities, and the worship of Vesta, her Roman counterpart, became a state cult (Fig. 2.7), best known for the vestal virgins who tended the perpetual flame in the shrine of Vesta. Use of fire in spiritual or magical rituals has continued throughout the world. In Europe, fire festivals (based on bonfires rather than landscape burning) were common since pre-modern times (the Middle Ages, AD 410 to the1450s) until the nineteenth century. These are exemplified by springtime fertility festivals, such as the Lenten Fires of North-West Europe and the Beltane Fires of the UK Celts complete with human sacrifices, the midsummer fires at the summer solstice, and their counterpart at the winter solstice (which commonly involved the Yule log that we still remember at Christmas) to ensure prosperity and safety. Over the same time span, people in many parts of Europe have resorted to ritual fires – 'need-fires' – at times of great distress or calamity, such as their cattle being attacked by epidemic disease.

Figure 2.7
Remains of the Vesta temple in the Roman city of Ephesus, now in modern Turkey.

useful for killing and cooking small mammals and insects such as grasshoppers, and for improving access to edible plant roots and bulbs. A useful summary of the use of fire in North America can be found in Mellars (1976).

Fire was also useful in removing accumulated vegetation, making travel easier and creating fire breaks around camps. It is difficult to argue with the fact that the value of fire in maintaining biodiversity was clearly recognised by early people around the world. The Aboriginals of Australia certainly used fire in a big way to 'clean up the country' – returning the land to an orderly state and cleansing it of evil spirits using 'corrective fires'. Stephen Pyne (1993) rightly suggests that 'Wherever humans and ecosystem have co-evolved, not to burn could be as irresponsible as improper burning. Good citizens used fire well; bad ones, poorly, or not at all.' This illustrates the important influence people had over the vegetation of the land they lived on.

As early humans learnt to use fire, they had at their fingertips a tremendous force for modifying the landscape around them, helping pave the way for a transition from being hunter/gatherers to being more agrarian and settled people. But how carefully did they use fire?

Control of fire – careful or careless?

There has been a long debate about just how carefully aboriginal people used fire. Some have argued that these people were mainly concerned with their immediate survival with little concern for the long term, and so were careless with fire. Either that or they lacked the skill to adequately control the fires they started. After all, fire is the one major ecological force that can be very easily unleashed by humans, with a huge potential impact on the landscape. This view is bolstered by the many European observers who remarked on the casualness of fire use by North American aboriginals, fires being left to 'burn out of control' and abandoned campfires left still burning after an overnight stop. Early visitors to Australia, the tropics and many other parts of the world in the eighteenth and nineteenth centuries similarly record frequent dense smoke and roaring fires that seemed to cover the whole land as the Aborigines burned with abandon.

Others – the majority – would argue that hidden in this apparently careless attitude is tremendous knowledge and understanding, in the same way that a concert pianist makes the playing of a complex piece look easy and effortless. Jones (1975), looking at aboriginal use of fire, concluded that 'People used fire accurately, aiming them into a natural break such as an old fire scar or swamp, timing the fire so the predictable wind changes later in the day would blow them back into their own track, or so that the evening dew would dampen them down.' On other continents there is also abundant evidence that the various aboriginal people have been (and are) remarkably skilled in fire management and just 'knew' where and when to burn. They knew how to take advantage of the benefits of fire while trying to avoid the dangers. Stephen Pyne's many excellent fire history books (Pyne 1995, 1997, 2001a, 2001b, 2004, 2007) give much more detail on this fascinating topic.

Aboriginal peoples could carry fire with them as they travelled. Those of North America carried fire brands or 'slow matches' – a tightly rolled rope of bark smouldering at one end, and the Aboriginals of Australia had the infamous and valuable 'firestick' – a slow-burning piece of wood or bark. They could also start fire from scratch by a variety of methods. With fire being easily started and transported, the question that arises is just how widespread were the fires of indigenous people: did their fires affect large areas of landscape or was this localised around their settlements; and did they significantly alter the frequency of fire over large areas?

Having said all this, it is important to recognise that not all aboriginal people used fire to modify the landscape. Natcher and colleagues (2007) showed that the Koyukon people of the western Interior of Alaska 'considered fire a destructive force and had no recollection or oral history of using fire for landscape management'. In contrast, the Gwich'in of the eastern Interior actively used fires to manage the landscape. As the authors put it: 'These findings call into question the commonly held view that the native peoples of North America pervasively and near universally modified landscapes through the use of fire.'

Effect of aboriginal people on the landscape

In areas of the world where fire has been fairly rare since the last Ice Age, such as northern Sweden (see Carcaillet *et al.* 2007), it is clear that humans had little impact on the naturally occurring cycle of fire until the last two centuries. But in more fire-prone parts of the world, the picture appears to be very different. Nevertheless, it has been a long-held view that even in fire-prone areas early aboriginal people were too few to greatly influence vegetation on a large scale (Russell 1983, Parshall *et al.* 2003). Williams suggests in his book *Deforesting the Earth* (2003) that this partly stems from the fashionable view put forward by early writers in North America who 'extolled the idea of an untouched, virgin and virtually uninhabited wilderness' against which the first European settlers battled. This image of heroic struggle has 'become a part of the American heritage' and difficult to argue against. In this vision, the aboriginal people lived in harmony with the forest, creating a barely perceptible human disturbance, and it was the 'blundering European who transformed, destroyed, and devastated it all'.

Given the skills that early people developed over millennia in manipulating fire (or if you prefer, the ease with which they could ignite uncontrolled fires) the proposition that the arrival of people in fire-prone areas resulted in significantly more fire seems quite realistic. There was certainly more fire around their habitations. For example, in eastern North America, the aboriginals cleared areas that are recorded as being handfuls to tens of kilometres in width. And it is highly likely (though not universally agreed) that their influence on the landscape was much wider, in much the same way that the Aboriginals of Australia covered the whole landscape with fire. Fires in Minnesota conifer forests were huge prior to European settlement, with a mean size of 4000 ha and a maximum estimated size of 160 000 ha. Moreover, it is suggested that by the time that Europeans arrived in North America, the aboriginal peoples may have as much as doubled the number of fires that would normally have burned due to lightning. Indeed, the Maskouten Indians of Wisconsin used fire so often that Father Pere Marquette called them the

'Fire People' when he met them in 1673. However, the unsupported assertion some-times made that such people burnt every acre of the west does appear exaggerated.

It would seem, then, that in easily burnt forests, pre-European human-caused fires (deliberate and accidental), coupled with natural fires, resulted in widespread alteration to the landscape, changing the density and species of trees, and encour-aging more open areas over large areas. Archaeological remains of the Iroquois in eastern North America are associated with high charcoal levels and a change from beech and sugar maple (*Acer saccharum*) to oaks and eastern white pine (*Pinus strobus*). The importance of human fire in such changes would obviously have varied; much less important in areas with few humans and more important in areas of high habitation such as along the coastlines and river valleys of eastern North America.

There are very impressive examples of dramatic change to the landscape by burning. For example, giant marsupial lions and kangaroos that stood 3 m high and weighed over 200 kg, plus other Australian giants that lived 200–400 000 years ago, had been thought to have become extinct due to arid conditions during the approach of the last Ice Age. However, these extinctions are now thought to be due to the extensive fires started by the early humans who changed the landscape so much that large herbivores ran out of food, so starving the large carnivores (Pri-deaux *et al.* 2007). A more recent example stems from when the Maoris reached New Zealand in the late thirteenth century. They lived mostly on the moas, a group of large flightless birds that ranged in size from a small goat to giant ostrich-like beasts up to 4 m tall, weighing 250 kg (over twice the size of an ostrich). The birds provided food, clothing and most of their implements – a truly important group. The moas were hunted with the aid of fire, and very successfully because it is estimated that within 120 years the Maoris had reduced an estimated population of over 150 000 birds to eventual extinction. Moreover, the population of around 10 000 people in the South Island had removed at least 3 million hectares (8 million acres) of forest, primarily by fire (see Anderson 1989, Holdaway & Jacomb 2000, Williams 2003 for more details). A similar example is seen in Patagonia in South America. After the late nineteenth century wars decimated the fire-using native population, the once open grassy steppes were readily invaded by *Austrocedrus* (Chilean incense cedar) of ever-increasing density and area (see Alaback *et al.* 2003). This does underline the point that aboriginal peoples' use of fire could bring about large changes in the landscape, which today we would look upon as irresponsible and ecologically unsound. This is not to say that they were necessarily careless, unintelligent or unaware, and before we think harshly of our ancestors, we should pause and think of how we have manipulated the landscape for our own use (creating, for example, extensive farmland), and our current problems over climate change. Nor is this picture of great change always true; the Malagasy people of Madagascar took great care to conserve their forests unchanged.

As well as introducing more fire into the landscape, aboriginal peoples also altered the type of fire. An example is to look at the people who settled amongst the coastal redwoods (*Sequoia sempervirens*) of California about 12–11 000 years ago bringing fire with them. Fire frequency (see Chapter 4 for a review of what this means) went from one fire in 135 years to one fire in 17–82 years and in some areas every 8 years. Due to the consequent reductions in the build-up of fuel, the fires changed from infrequent, large and intense surface and crown fires (again see Chapter 4) to fires of lower intensity with just periodic high-intensity surface fires. Fortunately for us and tourism, the change favoured the spread of the coastal redwoods and stopped Douglas fir (*Pseudotsuga menziesii*) and western hemlock (*Tsuga heterophylla*) from taking over (we have a lot to thank these people for!). This pattern of aboriginal peoples being implicated in more frequent but less severe fires is fairly universal.

It can be difficult to disentangle the effect of humans on fire in the landscape from natural changes due to climate variation because of the way these two factors affect each other. Climate variation will have had some influence on human activity, and large-scale human impact on vegetation has in turn led to climate change in at least some parts of the world. Perhaps the best example of this is in Australia. Humans are known to have occupied much of the Australian continent for at least the last 40 000 years (and possibly much longer) and they undoubtedly brought with them a good knowledge of fire-making and its uses (see Pyne 1991). On this driest continent (after Antarctica), Aborigines made a large impact on the vegetation of the whole continent within a short period of time. In the fire records (see Chapter 4 for a discussion on evidence of prehistoric fires) there is a surge in charcoal peaks and associated vegetation changes from around 40 000 years ago onwards. However, it has been said that Aboriginal burning just accelerated an existing trend rather than starting it, with climate change being the real driving force behind the drastic changes in vegetation. In this case, climate variation increased the effect of human burning, and the burning, dramatically altering the vegetation, undoubtedly helped to speed the change in climate.

Arrival of the Europeans

Beginning with the Renaissance voyages of discovery, Europeans spread around the world. They took with them the fire culture of their home, that is, a view that fire was not 'natural' in terms of being beneficial, but was paradoxically a useful if coarse tool for making the landscape more useable. Indeed, while fire suppression resulted in fewer fires in Europe in the early to mid nineteenth century, European settlers around the world used fire in a cavalier fashion, clearing land with fires that sometimes developed into very large fires. In North America, the biggest recorded fire – the Miramichi Fire of October 1825 – burnt 1.2 million ha (3 million acres) of New Brunswick, Canada and Maine, USA, killing more than 160 people and making 2000 homeless (see Pyne 1982, 2007 for a more detailed description). It started as a series of small logging and settler fires during a summer of severe droughts and high winds. The fires coalesced into one huge complex fire, made worse by the large amounts of dried and flammable logging debris. Similarly, the Peshtigo Fire in the Lake States region burnt about 500 000 ha (1.2 million acres) during October 1871, even though the main area burnt was (usually) non-flammable hardwood forest. This was helped by five dry months beforehand but the primary cause was the enormous amount of logging debris left from land clearance and logging. Felling timber was, of course, a necessity, a means of survival, producing building materials and fuel for domestic fires. In the early 1700s it has been estimated that a typical family in New England would use 20–30 cords of wood a year (a cord is a stack of cut wood $4 \times 4 \times 8$ feet, equivalent to 128 cubic feet or 3.625 m^3).

North America was not the only place where settlers played havoc with fire. In the early 1400s, the lush island of Madeira was colonised by Europeans and cleared for sugar cane production by fire. So great was the initial fire that the settlers took refuge in the sea 'where they stayed for two days and two nights up to their necks until the blaze burned itself out' (Williams 2003). Plus, between 1750 and 1900, 56 million ha of forest in the tropics had been converted to grassland (54% in Africa, 45% in Latin America and a small fraction in South-East Asia). A certain amount of this may well have been marginal forest and the use of fire just tipped the balance or accelerated its change to grassland – but still an incredible achievement.

In Australia, where Cook reported on the abundant use of fire by the Aboriginals, fire became even more widespread during the early years of European settlement to levels higher than at any other time during the last 10 000 years (i.e. since the end of the last Ice Age). By 1850, the state of Victoria was mostly pasture land with a European population of 77 000 – ten times that of the original Aboriginal population, and fire was used aggressively to clear land and encourage grazing. But catastrophes like Black Thursday (6 February 1851) – when fires burnt an estimated 5 million ha (12 million acres) – began to make people wary of the use of fire. But ironically, as burning declined and scrub grew, fire danger increased. This was followed by Red Tuesday (1 February 1898) when fires burnt 260 000 ha (600 000 acres) at the end of a dry summer.

These huge early fires were, though, an upward blip due to lack of care and control. The underlying feeling by European colonists worldwide, aided no doubt by the tragedy of these large fires, was that fire was a bad thing, and an alien phenomenon in the landscape. Thus, this mindset strongly influenced fire-control policies; fire should be suppressed and, if possible, completely prevented and removed as an unwelcome menace.

Why were Europeans so anti-fire? A number of reasons are put forward by Stephen Pyne (1995, 1997, 2001a). First, European agriculture had become intensive, organised, sedentary and fire-less. This was partly pressure of space but also due to politics – political power and taxation required an organised sedentary population whereas fire-based cultures tend to be nomadic. Consequently, fire was no longer a necessary agent to staying alive; it was a threat to wooden buildings and people's lives. Fire was also seen as a punishment from God – what awaited people in hell. Some thought that the Great Fire of London, 1–6 September 1666, was God's punishment for gluttony since it started in Pudding Lane and finished at Pie Corner. Moreover, forestry in Europe was growing as an organised business, where again the view of fire was that of a destructive agent. As a result of all these influences, colonists took their fire-less culture with them as an ideal to be aimed at. Fire suppression was also a powerful way of controlling indigenous people by taking away from them their traditional control over the landscape.

Accordingly, in the first half of the twentieth century, thoughts and activities around the world turned to 'fire control' to try and remove the destructive fire – that is, aiming at total fire exclusion. Feelings such as the following, voiced in Australia, were common to controlling fire:

> It is the first duty of a forest officer to study the best means of fighting the greatest enemy the forests have to encounter. The fire fiend is easily raised, and once started on its destructive career is very difficult to restrain. (Perrin 1890)

In the early 1900s the thinking and phraseology of fire was pervaded with words such as 'dangerous and destructive' and burnt areas were 'dark, dismal and desolate'; fire in forestry revolved around the 'five Ds' – devastation, depletion, deterioration, decay and disappearance.

Bambi and Smokey Bear

In North America, the anti-fire feelings were fed by two powerful developments during World War II. In the spring of 1942 (just after Pearl Harbor), a Japanese submarine shelled an oil field along the Californian coast, near Santa Barbara. This led to widespread concern that west coast forests were very vulnerable to incendiary attack. Moreover, resulting fires would be difficult to fight with so many people and resources tied up in the war effort. So a campaign was launched to reduce this risk, part of which was to urge people to be more careful with fire themselves. The Forest Service organised the Cooperative Forest Fire Prevention Campaign, which was responsible for slogans such as 'Forest Fires Aid the Enemy' and 'Our Carelessness, Their Secret Weapon.'

Around the same time in 1944, Walt Disney produced his now famous film, *Bambi*. Disney was equally worried about the west coast and because of this it is said that he deliberately built an anti-fire theme into the film. As you may recall, in the film Bambi's mother is shot by the nasty hunters. But that is not the worst. The foolish hunters allow their campfire to escape and poor little Bambi is threatened by towering flames, and it is only his majestic father arriving at the eleventh hour that saves Bambi from a certain and gruesome death. The real villain in the film is the fire – a deliberate and memorable anti-fire message.

Disney allowed the Cooperative Forest Fire Prevention Campaign to use Bambi on one of their posters. It was a great success but because Bambi was on loan from Walt Disney studios (in that instance) for just one year the campaign looked for another animal symbol. They settled on a bear – and so the famous symbol of Smokey Bear was born. The first poster featuring Smokey appeared in 1946 and since then Smokey Bear has blossomed into a national symbol recognisable by all Americans and identified with one message: 'Only <u>you</u> can prevent forest fires'.

These two icons of anti-fire, Bambi and particularly Smokey Bear, have much to answer for instilling concern about unwanted fire in many Americans. It is suggested by Ling & Storer (1990) that by 1973 Smokey's efforts had saved timber worth over 17 billion dollars in the National Park system alone (that was in the days when felling was permitted in National Parks). Smokey Bear's message underwent a small change in 2000, altering 'forest fires' to 'wildfires' (Fig. 2.8). As more people started living in flammable areas (the wildland–urban interface – see Chapter 6) it was felt that while 'forest fires' implied fires in the deep forest, 'wildfires' better communicated the message that unwanted fires can occur much closer to home.

Reduced fire frequencies

The overall effect of these fire-control policies was to reduce the amount of fire on the landscape in the mid twentieth century compared with before the colonial invasion (see Fig. 2.9). This resulted in wholesale invasion of scrub and trees to fill what had become open landscapes under aboriginal peoples' and early colonial care. These changes have been seen around the world from South America to Australia, and North America to Russia and Africa, and a good example in western Canada is explored in Box 2.3.

It can be seen that in many areas of the world fire suppression activities have been extensively aided by changes in human use of the landscape. In Scandinavia, for example, the decline of slash-and-burn agriculture in the nineteenth century, together with increasingly organised forestry activity (and thus firmer control over use of the forests), increased grazing and firewood collection (especially the felling of dead standing trees which are important for lightning ignition – see Chapter 4), aided by deliberate fire suppression activities, has resulted in a reduction in the amount of fire. In the Great Lakes area of North America, flammable conifer forests were converted to smaller, less burnable deciduous forests and to farms, cities and roads. This fragmented the burnable landscape and disrupted the flow of fire such that fire now plays a minor role in Great Lakes forests. Thus, forest fragmentation and urbanisation of the remaining fragments by, for example, the removal of dead wood for visitor safety and aesthetic reasons tend to lead to less fire. But reduced natural grazing of the flammable understorey grasses and shrubs, an increase in the number of people and hence arson (see Chapter 4), and the

Figure 2.8
The poster at the top is from 1954. Below, a poster from 2001 after Smokey Bear's message was changed in April 2000 with a subtly new message: 'forest fires' changed to 'wildfires'. Smokey Bear images used with the permission of the US Forest Service.

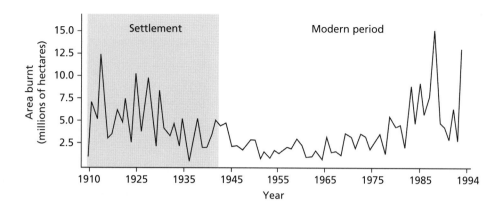

Figure 2.9
Average area burnt each year in western USA
between 1916 and 1994. Data complied by
Interagency Fire Center, Boise, Idaho and the US
Forest Service. Data prior to 1931 do not include
National Park or Indian reservation lands and data
prior to 1926 include only forested land, so the
actual burnt area is underestimated. Redrawn
from Alaback et al. (2003), Fig. 4.3 with kind
permission of Springer Science + Business Media.

effects of a changing climate (see Chapter 8) could shift the balance to an increase once more in the amount of fire in the landscape.

Fire control to fire management

After decades of fire control in the last century, the gradual change in many of the world's flammable forests, becoming denser with more dead material, with fewer plants and animals in the gathering shade, was becoming blatantly obvious. This began to worry a good many people in different countries who voiced the opinion that fire was an essential and natural ecological factor (see Chapter 5 for the arguments), and should be used as a positive tool. They had an uphill struggle because successful fire control meant that many influential people had less experience with fire except for those huge terrifying fires that broke out in the larger fuel deposits that were building up – fires which were driven by extreme weather and so impossible to do much about. This just served to reinforce in their minds that fire was a destructive agent that should be eradicated. Even in Australia, where it was appreciated that small frequent fires helped prevent the immense conflagrations such as Black Thursday, it was a difficult battle that resulted in awkward compromises.

Nevertheless, beginning in the second half of the last century, different countries began to change their emphasis from fire control (put all fires out) to fire management (using fire as a tool). Park managers especially set about restoring the pre-European fire regime. In places like India, Australia and South Africa where it was broadly known that fire was indispensable, it was gradually adopted as a management tool. It took longer in North America to realise that controlled burning in the guise of deliberate burning or allowing some fires to burn was an ecological necessity. Perhaps because of that, American and Australian fire policy has alternately swung between pro-fire and anti-fire but out of phase with each other.

Even when fire was considered by the controlling authorities to be necessary in the forest, it was not all clear sailing. There were plenty of challenges over the sorts of fires that should be allowed. A park manager might want a variety of fires,

Figure 2.10 (Opposite page)
Athabasca River valley in Jasper National Park,
Canada, photographed in 1915 by M.P. Bridgland
(top) and from the same viewpoint in 1998 by
Jeanine Rhemtulla and Eric Higgs (bottom).
Photographs are a mosaic of four photographs
taken from survey station No. 57 (photograph
Nos. 459–462). From Rhemtulla et al. (2002). ©
2008 NRC Canada or its licensors. Reproduced
with permission.

Box 2.3. Changes in Jasper Valley, Canada

Van Wagner and colleagues published a paper in 2006 that showed that in the Rocky Mountain parks east of the continental divide (primarily Banff and Jasper National Parks), there has been a distinct reduction in the fire cycle (the average interval between fires at any given point – see Chapter 4 for definitions) over recent centuries. The fire cycle between 1280 and 1760 was estimated at 60–70 years. After this, with the disruption of the aboriginals' traditional use of fire, the area burnt each decade became rather erratic but overall was greatly reduced so that by 1940 the cycle had increased to around 175 years. From 1940 to 2000 the area burnt sharply decreased still further. Various reasons for this decline were explored, including changes in weather patterns, but it seems that improved fire control was the primary reason for the reduction in fire. Fire suppression began around the town of Jasper in 1913 and the present forests originated primarily after large fires in 1889, 1847 and 1785, before fire suppression efforts began.

It is obvious from the photographs in Fig. 2.10 that there are more trees in the valley following fire suppression. Between the photographs the area of forest increased from 35% to 65% with an increase in dominance of conifers over aspen, and grassland area decreased by almost 50% (Rhemtulla *et al.* 2002). This has been taken as clear evidence that human fire suppression activities have had a large impact in reducing the amount of fire ('fire protection this century may be responsible for increasing fire cycles' – Wierzchowski *et al.* 2002) allowing trees to invade. But it could also be argued that declining densities of large grazing animals, such as elk and deer, by hunting and habitat removal, is as (or more) important in allowing more young trees to grow unmolested.

But does fire suppression really reduce the amount of fire? It is argued by Ed Johnson (see his 1992 book and Johnson *et al.* 2001) that fire suppression has played a minor role in reducing fire occurrence over the last century in North American conifer forests. He argues that 97% of the landscape is burnt by the largest 3% of fires and these are determined by large-scale dry weather, making the fires impossible to control, even by fire breaks. Fire suppression can put out the other 97% of fires but as these make up only 3% of the area burnt, it makes little difference to area burnt each year. Others have argued (e.g. Lorimer & Gough 1988) that smouldering 'holdover' fires in wet periods that are now put out would naturally have been the cause of many a large fire once the weather was dry enough; therefore fewer large fires now start. However, Johnson counters (see Johnson & Gutsell 1994) that the area burnt per year in remote areas with no suppression is similar to those areas with active suppression. David Martell gives evidence that disputes this (Martell 2002), and in turn Johnson's group dispute Martell's evidence (Miyanishi *et al.* 2002). The argument continues.

M.P. Bridgland 1915

J.Rhemtulla and E. Higgs 1998

including large intense fires, to maintain a 'natural' landscape with all its plants and animals, while a nearby property owner would want frequent low-intensity fires to keep fuel levels low without endangering their property. Or members of the public, looking at things through their cultural eyes, might see all fires as unnatural and an ecological disaster and challenge loudly whether fire is needed at all.

Part of this problem is that although fire is an ecological necessity in wild forests, it is still a dangerous enemy when in the wrong place at the wrong time. Socially and economically it can be devastating if, for example, the wood supply to a saw-mill is destroyed by fire (the trees will regrow but how does the mill survive in the meantime?). This helps explain why the Food and Agriculture Organization (FAO) still despatches fire experts to Mauritania, Nepal and Patagonia to advise on forestry and fire suppression.

Even where fire would be beneficial, it may not be practical. The 1995–2005 management plan for the Blackwood of Rannoch, a remnant of native Scots pine forest in Perthshire, Scotland, states that

> Although fire can be a natural process, the isolation of the Black Wood from
> other areas of native pinewood means that any fire outbreak will be brought
> under control as quickly as possible. This will require the use of helicopters
> and foam application in any significant fires. Ground teams with beaters
> and water pumps will also be used.

In this case because the whole forest is only around 1000 ha – a remnant of a once much larger forest – a single fire could sweep across the whole forest, wiping out whole populations of animals and possibly plants which would have nowhere left to re-invade from.

Are fire frequencies increasing again?

Fire appears to be on the increase once more. The USA had its worst fire year in 50 years in 2006 (96 385 fires burnt 9 873 745 acres or 3.9 million ha) closely followed by 2007 (85 822 fires burnt 9 321 326 acres or 3.7 million ha). Australia has also experienced more bad fire years lately: in the state of Victoria 1.3 million ha burnt in the summer of 2002/3, the largest area since 1939; the 2006/7 fire season was also bad following 50 rainless days; and the season of 2007/8 was even worse. Even the UK had extensive fire problems in the early spring of 2003, and Portugal and Greece had record areas burnt in 2005 and 2007, respectively. Inexorably, the area of land burnt appears to be creeping up in many places around the world.

Part of this can be attributed to drought – either long term over several years as in Australia or short term as in the dry 2003 spring in the UK. Whether this is due to climate change is explored further in Chapter 8. But a clear additional factor is the build-up of fuel due to fire suppression activities. The extra fuel helps an initial spark to develop into a fire, and helps the fire to grow quickly in size and intensity to a level that is hard to control, resulting in larger and more intense fires. A vivid example of fuel build-up can be seen in the ponderosa pine (*Pinus ponderosa*) forests of western North America. These forests evolved with low-intensity fires creeping through at regular intervals (every 2–15 years). With fire suppression, the forest has acquired a build-up of litter and shade-tolerant trees. One study of northern Arizona forests found that the present-day density of trees is 40 times that of pre-settlement times. But we should not automatically conclude that fuel build-up is always the culprit in severe fires. The knee-jerk reaction after the 1988 fires of Yellowstone National Park which burnt an area of 730 000 ha/296 000 acres (the largest since a 'natural burn policy' was instituted in 1972) was to attribute their fierceness to the large fuel accumulation due to 50 years of fire suppression. This undoubtedly contributed but the main factor was unusual summer drought and strong winds on top of a steadily changing climate (see Alaback *et al.* 2003 for a detailed study of the Yellowstone problem).

Even wet tropical forests are seeing more fire due to drought (caused by El Niño – see Chapter 4). They are also becoming richer in dry burnable fuel, partly due to drought but also due to forest clearance and logging, leaving more dead debris to burn. Fires starting in clearings are also becoming intense enough to be able to burn into the surrounding undisturbed forest. This usually means that successive fires burn further and longer, spreading faster and higher into the tree canopy. It is food for thought that satellite analysis of the Brazilian Amazon showed that in 2007 between 50 700 and 72 300 fires burnt (depending upon the satellite used), the highest on record since 2003, and over recent years something like 20 000 to 40 000 km^2 has been burnt in the whole Amazon Basin per year.

3 | How a fire burns

Fire is a chemical reaction that results in the very rapid release of the energy stored in fuel. Plants use the sun's energy to combine carbon dioxide with water to produce carbon compounds (sugars, starch, cellulose etc.) and oxygen. When a plant dies and decomposes, the reverse happens: oxygen is used up in breaking apart the carbon compounds to release carbon dioxide, water and a gentle trickle of heat (think about the heat produced by a good rotting compost heap or pile of manure). Fire is a form of decomposition, just a lot quicker, releasing the carbon dioxide, water and heat in a massive burst. In this chapter we start at the beginning and look at how plant material burns before considering how a whole fire burns.

Mechanics of fire

When heat is applied to, for example, a piece of wood, three distinct phases are passed through: pre-ignition, ignition and then combustion.

Pre-ignition

When heat is first applied to our piece of wood, the energy will initially be absorbed by water contained in the wood (which has a large capacity to store heat). Then as the temperature rises further (Fig. 3.1a), the heat evaporates the water at the surface of the wood (it may even steam as the water vapour condenses in the air) which keeps the wood surface around 100 °C (see Box 3.1 for a demonstration of this). The water is thus driven away from the surface of the wood – the 'combustion zone' – by moving both into the air and deeper into the wood (and so increasing the moisture content deeper down). When the wood surface is dry, its temperature is no longer maintained at 100 °C by water evaporation, and as a result it will get hotter and other substances in the wood such as terpenes (these give the piney smell of woods), fats, oils and resins start to evaporate. These form a cloud of flammable gases over the fuel (Fig. 3.1b). When the wood reaches around 130–190 °C it starts to break down chemically and over 260 °C the cellulose in the wood starts to scorch, producing a blackened surface of char (almost pure carbon) and a grey smoke of wood alcohol and tar gases. Lignin in the wood also joins in but

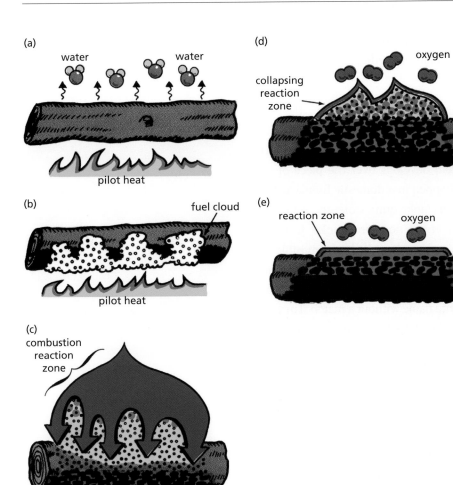

(a) water   water

pilot heat

(b)   fuel cloud

pilot heat

(c) combustion reaction zone

charred fuel

(d) collapsing reaction zone   oxygen

(e) reaction zone   oxygen

Figure 3.1
Stages in the burning of a piece of wood. (a) Heat applied from the outside (pilot heat) causes water to evaporate then (b) the fuel is chemically decomposed to form a fuel gas cloud. (c) When this is ignited by the pilot heat (see the text for other ways in which this can happen) a flame is formed which burns at the junction of the fuel gas and oxygen (the combustion reaction zone). As the fuel is used up (d) the supply of flammable gases gets less and char (unburnt carbon) builds up sealing the gases in. Eventually the flame goes out and (e) the remaining fuel burns by glowing combustion without a flame. From Cottrell (2004).

Box 3.1. Setting fire to a bank note without burning it

If you hold a match to a bank note it will obviously burn, being made of paper. The heat of the match will raise the temperature of the paper past its ignition temperature until it catches fire. Similarly, if you soak the note in pure alcohol, the note and the alcohol will both rapidly go up in flames since both are flammable. However, try soaking the note in a mix of water and alcohol. When you hold a match to it, a flame burns brightly which eventually goes out leaving you with a wet and warm but unburnt note. Why? The key comes from the large heat capacity of water.

The flame comes from the alcohol; as it burns the heat is absorbed by the water and, as it gets warmer, is used in evaporating some water. All this keeps the paper of the note below its ignition temperature until all the alcohol is used up and the flame goes out. The alcohol and water mix needs to be just right, usually around equal volumes of each. Too much water, no flames; too little water and it will all be evaporated before the flame goes out and the paper will catch fire. Try experimenting with strips of paper until you've got it right!

needs a higher temperature of 280–500 °C, producing fewer gases and more char than cellulose. The scorching is just like burning the toast – you get blackened bread from scorching and a puff of smoke but no flames (unless you go to the ignition phase). Tars in gas and condensed droplets contribute greatly to the build-up of flammable gases around the wood. This breakdown of the fuel to flammable gases is called pyrolysis (pyro-lysis, heat-divided).

Ignition

If a pilot flame touches the flammable cloud it will be ignited and a self-supporting flame (i.e. the pilot flame is no longer needed) develops over the piece of wood (Fig. 3.1c, d). Sudden ignition can be seen to happen in a domestic fireplace. If new wood is added to glowing embers it will start to smoke quite violently; if a match is thrown in, the smoke erupts in flame and you see no more smoke, just flames because the only things left are water and carbon dioxide, both invisible. This is the cloud of pyrolysed gases catching fire. If you wait longer before applying a match and the temperature gets higher, the unpiloted (or spontaneous) ignition temperature is reached and the smoke bursts into flame without a match. The piloted ignition temperature of gases from wood is around 320–350 °C, but the 'flash-point' for unpiloted ignition is about 600 °C (Babrauskas 2002).

Combustion

The radiant heat from the flaming combustion continues the process of pyrolysis, boiling off more short carbon-chain fragments of gas which enter the fuel cloud and eventually burn. The radiant heat also preheats and ignites unburnt wood, and the wood is consumed by the fire in a run-away process.

In this flaming combustion it is not the wood itself that is burning but the gases released by pyrolysis that burn just above the wood. The flame itself consists of a central core of unburnt gases surrounded by a thin burning envelope (the 'reaction zone') where enough oxygen mixes with the gases to allow combustion. The hotter the fire the more the quantity of flammable gas being produced, the further it will travel before mixing with enough oxygen to burn, so the longer the resulting flame. It is also why a puff of wind over a campfire results in suddenly smaller flames – more oxygen is mixed into the gases lower down near the fuel and it burns there instead. It is clear from this that three things are needed for a flame: fuel, heat and oxygen (this is the fire triangle described in Box 3.2).

By-products of the flame are primarily carbon dioxide and water. It may seem surprising that fire produces water but burning is a simple reversal of photo-synthesis in plants (see above). In fact burning 1 kg of dry plant material produces around 0.56 kg of water plus any moisture that was in the fuel at the time of burning, which explains why something like a paraffin/kerosene stove which vents into the room (rather than a chimney) will steam up the windows. In a forest fire the water vapour billowing around can be important as a dilutant of the combustible gases.

Box 3.2. The fire triangle

The fire triangle has been used to describe the three important constituents of fire – fuel, heat and oxygen. You can see from this description that if you take any one of these away the fire cannot burn. The most basic principle of firefighting is to remove one or more of these elements. From Cottrell (2004).

Normally 50–95% of the carbon released from the fuel is as carbon dioxide but the remainder goes to make up the constituents of smoke, a complex mixture of gases and sooty particles mixed with water vapour. Soot is made of fragments of carbon, tar and other chemicals that clump together or around tiny fragments of unburnt fuel and ash to form less-burnable lumps. These clumps absorb heat from the surrounding reaction zone and glow, giving the orange to red colour of the flame centre. Soot particles that escape unburnt above the flame rapidly cool, giving the black colour to smoke above the fire. Generally, more vigorous fires produce more smoke because it seems that with bigger flames there is more turbulent airflow and more pockets of gas and soot escape unburnt above the flames. Soot particles need to be exposed to temperatures above 800 °C with high oxygen levels to be effectively burnt and so can easily escape burning.

As the wood burns, charred carbon (char) and ash builds up in a layer covering the wood surface and although pyrolysis is still going on, it is not sufficient to support the flames (Fig. 3.1d). At this point the flames die down and flaming combustion ends. But if the surface of the wood is above 500–600 °C and oxygen can reach the surface (volatile gases and flames may prevent this before) the carbon itself starts burning by smouldering or glowing combustion (Fig. 3.1e).

In glowing combustion, it is the remaining solid carbon that burns rather than gases burning above the wood. Heat continues to break down the carbon chains in the charred wood, which mix with oxygen at the charred surface to burn with an orange glow rather than with flames. If the supply of oxygen is limited, carbon monoxide rather than carbon dioxide is produced inside the fuel. As this escapes and mixes with more oxygen, it is burnt to produce carbon dioxide, producing a very small blue flame. Glowing combustion is just the way that charcoal in a barbecue burns – a hot glow but no smoke or flames. Charcoal for a barbecue is produced by heating wood in an atmosphere largely devoid of oxygen to drive off the volatile gases without allowing ignition. The solid fuel coke is produced in the same way from coal for use in urban areas where smoke is unwanted.

Although glowing combustion itself produces little visible smoke, you will see some smoke if some of the log is unburnt. The energy of the glowing combustion heats the adjacent unburnt wood, releasing grey tarry smoke (the pyrolysed but unignited gas cloud). Usually there is not enough heat in the glowing combustion to ignite the fuel cloud (but see the section below on Ground Fires). This is why smouldering logs in a fireplace result in tar gases which condense inside the chimney forming a dangerous flammable layer which can cause a chimney fire if they ever catch light. Indeed, glowing combustion (in both wood and the organic matter on the soil) can produce significantly more soot and tar (i.e. smoke), along with such gases as carbon monoxide and methane than the flaming part of a fire where they are destroyed by burning. Ironically then, low smouldering fires and really big flaming fires produce more smoke than intermediate fires.

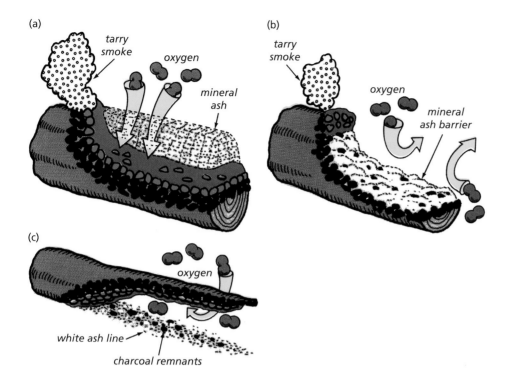

(a)

tarry smoke

oxygen

mineral ash

(b)

tarry smoke

oxygen

mineral ash barrier

(c)

oxygen

white ash line

charcoal remnants

Figure 3.2
As the glowing combustion in Fig. 3.1e carries on, ash gradually accumulates (a, b diagrams) and creates a barrier to oxygen reaching the combustion zone, and so the fire goes out. But if the ash is removed or falls away (c) then combustion can continue until all the wood is consumed. From Cottrell (2004).

The relatively cool burning of glowing combustion (usually around 600 °C compared with around 800–1200 °C in a flame) converts the carbon in the fuel to carbon monoxide which is usually converted in turn to carbon dioxide in the oxygen-richer air above the glowing fuel. These two stages each produce about a third of the heat (i.e. two-thirds overall) as burning the same amount of fuel in flaming combustion. Glowing combustion thus has a relatively low heat yield.

As the wood burns, carbon is consumed as fast as char is formed and it should continue until all the wood is used up. Two main things will stop it. First, if part of the wood is wetter, the glowing combustion will go out if more heat is needed to evaporate the water than is generated by the combustion zone. Second, a layer of grey ash – the unburnable 'mineral' component of the wood that generally makes up less than 1–2% of the weight of wood (detailed below) – is left behind after combustion. The ash has no structural strength and collapses to form an ash barrier that, as it becomes thicker, progressively stops oxygen from getting to the glowing zone. At this point, starved of oxygen, the glowing combustion goes out, leaving a partially burnt log as in Fig. 3.2b. This is why stirring a fire helps it burn better, by knocking off the build-up of ash. This is also why fallen logs in a forest fire usually burn mostly from the bottom to the top since the ash falls away from the burning zone allowing it to continue (Fig. 3.2c).

As an aside, the charring and consequent slow combustion of wood means rather ironically that in a building fire, wooden beams remain more structurally sound that do metal ones. Metal beams lose about half their strength at 500 °C while wood loses very little strength at that temperature.

Heat transfer

To understand how fire spreads from a single burning piece of wood to many other pieces of fuel, the process of heat transfer becomes important. Once a piece of fuel is burning, for the fire to continue heat must be passed or transferred to the next yet-to-be-burnt pieces. Heat is transferred in three different ways from a flaming combustion zone; conduction, convection and radiation. Each of these processes has its own special little place in fire propagation.

If fuel particles touch then conduction of heat from piece to piece can help, but generally, wood and other plant materials are very poor conductors of heat and conduction is most important in heat penetrating into a large piece of fuel and fires burning below ground level. Very rotten wood has a low density and conducts heat even more poorly than sound wood. So when ignited the heat stays in one place, little flammable gas is produced from surrounding wood, any flames soon go out and it burns by glowing combustion. And this is why dry, rotten logs and stumps are a major player in spot fires (described later in this chapter) – embers that cannot ignite litter or sound wood (because the heat is too rapidly dispersed) will readily ignite the rotten wood.

Movement of hot air as it rises and is blown by the wind (convection) is more important in heating new pieces of fuel above the flame, possibly the crowns of trees above a surface fire, or fuels on steep ground. Convection of heat also plays an important role in very large fires where a plume or convection column forms (again, described later in this chapter). (As an aside, convection is indispensable to a flame. A candle will not burn for very long in space because without gravity there is no convection, and without convection, the carbon dioxide and water vapour produced by the flame stay around the flame and put it out.)

The last major way of heat spread is by radiation – the feeling of warmth on your face when standing in front of a fire which can be cut off by a hand in front of your face blocking off the radiation. Heat radiating from burning fuel is important in preheating fuel ahead of the fire. Radiation travels in all directions from a combustion zone, and as the fire develops from a single point (e.g. a campfire radiating heat in a circle) to a line (e.g. a forest-fire flame front) the effectiveness of the radiation changes. A point (or campfire) radiates energy in all directions, with the amount of energy received being dependent on the distance from the source. The relationship between distance and energy received is the inverse square of the distance; that is, as you double your distance from the fire, you receive one-quarter the energy. A line of fire also shows a reduction of radiant energy with distance, but the rate of reduction with distance is roughly equal – that is, each doubling of distance halves the energy received. Why is radiation important and considered the major force in governing fire spread rates? Because when wind tilts the flames over from their natural vertical angle, the radiant energy received by the fuel ahead of the fire greatly increases; this preheats the fuel much more effectively and speeds the spread of fire.

Heat can also be transferred directly by the movement of burning fuel, for example a burnt log rolling down a hill, or embers (bits of burning debris) carried aloft in hot air currents and deposited ahead of the fire front, sometimes literally miles away.

Variation in fuel quality

Flammability can be defined as the ease of burning. But what does this really mean? There are various views but here we will define flammability as ignitability or ignition delay time – the time taken from the start of heating to the point where the fuel is sufficiently chemically decomposed for ignition to occur. The readily flammable parts of a plant include live and dead leaves, twigs, bark and to a lesser extent wood. Different plants will have these in different proportions at different times of the year. Dead litter, twigs and branches too will accumulate underneath forest canopies at different rates and lie on the ground in different ways (e.g. the leaves of broadleaved trees like oak form a compact layer while needles from a pine tree form a more airy, less compact mat). The effect of the arrangement, amount and timing of fuel will be dealt with in the next chapter; here we will look at the factors affecting the flammability of individual bits of fuel.

Three main factors affect flammability: the moisture content of the fuel, its chemical make-up and its size and shape.

Moisture

The moisture contained within a piece of fuel is of overriding importance in that it affects all aspects of fire: whether the fuel will burn, the fire's speed in spreading, intensity, smoke production, fuel consumption and the mortality of live plants. In terms of flammability, a fire will only continue if there is enough heat to evaporate the moisture so that pyrolysis can happen; and evaporation is expensive in energy. For every gram of water evaporated from a piece of fuel, an extra 2260 joules of energy (540 calories) is required before ignition (termed the latent heat of vaporisation). The maximum moisture content that a fuel can have and still burn is referred to as the moisture of extinction. This is not fixed since fuels that burn more easily (for example if they have a high oil content, or are in smaller pieces) or are in a hotter fire either generate or receive more heat and thus will burn at a higher moisture content. Nevertheless, approximate figures can be used – see Box 3.3. Fuel moisture can vary from as little as 1–2% (Box 3.3 explains what this means) in dead litter in arid deserts to 300% in rotting wood and soil organic matter, and even 1000% in some living tissue.

Even below the moisture of extinction threshold, the amount of moisture still makes a difference to how things burn. In fine forest litter, fire spread slows as moisture increases between 0% and 20%. Between 20% and 34–40% the rate of spread decreases very little but at the upper limit it becomes zero and the fire goes out – the moisture of extinction.

Box 3.3. Moisture of extinction

Moisture of extinction for different fuels is the maximum amount of water that a fuel can hold and still burn. These are expressed as a percentage of dry weight. For example, 50% moisture means that for every 100 g of dry fuel, there are 50 g of water.

Grasses	12%
Dead forest fuels	25–49%
Soil organic matter	40–50%
Living fuels	120–160%
Resinous conifer needles	Nearly 200%

Chemical make-up

As a rule of thumb, plant tissue is composed of approximately 50% carbon, 45% oxygen and 5% hydrogen (by weight). Wood is made up of 41–53% cellulose (50–75% if you include the shorter chains of hemicelluloses) and 15–35% lignin. Lignin is harder to decompose than cellulose and so dead fuels become progressively higher in lignin; soft rotten ('punky') wood may be up to 75% lignin. The variations in composition reflect to some degree how well different fuels burn. However, two other small components of fuel are by far the most important – the quantity of 'extractives' and the amount of ash.

Extractives Extractible compounds (alcohols, fats, waxes, oils, gums, sugars etc. – called extractives because they can be chemically or physically extracted from plant material with ease) are a minor component of fuel weight making up just 0.27% of the total weight of some grasses to more than 15% in some pines and aromatic shrubs. But the extractives are generally very flammable and produce lots of heat when they burn (they have a high calorific value). Even small differences in extractives have a very important influence on the flammability of different species, and flammability of a species in different seasons. Many plants are most flammable (for a given amount of moisture) in early spring and summer due to variation in chemical composition.

Several examples illustrate the importance of extractives. In Californian chaparral, herbicides have been used in a set of experiments to kill the shrubs, allowing them to dry out and make them more flammable. It did not work, because although the foliage moisture fell from 114% to 8%, the extractives also fell from 16% to 3% and the treated shrubs were less flammable than untreated controls! Similarly, Malcolm Gill and Peter Moore (1996) tested the flammability of a wide range of cultivated plants in Australia in a furnace at 400 °C with a spark ignitor and found that some species (including two *Hakea* species and a *Podocarpus* species) were quick to ignite when fresh but slow after drying in an oven at 95 °C for 22 hours, presumably due to a loss of volatile oils.

For the same reason, the foliage of *Eucalyptus* species burns as well green as dry due to the high oil content. Webb (1968) remarked that fire damage may be increased by the explosive gases composed of *Eucalyptus* oils. Indeed, it was shown in the 1950s by McLaren (1959) that in Australia certain mixes of eucalyptus oil vapour and air form an explosive mix which, under special circumstances, could be responsible for forest fires erupting up to 10 km (6 miles) ahead of the main fire front, when combined with strong winds. Webb also noted that the leaves from *Eucalyptus* species in southern Australia, where crown fires are more common, are richer in more flammable oils with a lower boiling point than those of eucalypts farther north where crown fires are generally lacking.

Ash The ash content of plant material (what is left after everything combustible is gone – see Fig. 3.2) is less than 1–2% in wood but rises to 5–10% in foliage and even up to 40% in arid or harsh environments. The minerals in ash tend to reduce tar formation and increase char, and hence retard flammability and increase glowing combustion. It may be the phosphates that are mainly responsible for suppressing flammability (certainly phosphates are used in fire-retarding compounds). Ash content proportionately increases with decomposition as the easily rotted carbon in cellulose is broken down and removed. Thus, in forest litter in southern USA, ash content has been found to increase by a third after one year of weathering and the lowest litter layers can have 2–8 times the ash content of fresh litter. This goes some way to explaining why ground fires (see below) in well-rotted organic material accumulated at the soil surface tend to burn by glowing combustion. But, be warned, ash content does not always reduce flammability. As with many limiting factors, ash content is only limiting under certain ranges; when other factors increasing flammability are added up, they can override the retarding effect of the ash.

Size and shape

Size matters with fuel. Or perhaps not so much size as the surface area-to-volume ratio. That is, the amount of exposed surface area on a fuel piece in ratio to the volume of space that fuel piece occupies. Typically, smaller fuel pieces have higher surface area-to-volume ratios (see Box 3.4). Why is surface area-to-volume ratio important? Essentially for two reasons:

1. Remembering that burning or pyrolysis is chemically the oxidation of fuel, the higher the surface area-to-volume ratio the larger the relative amount of fuel exposed to heat and oxygen, providing more opportunity for the fuel to become involved.

2. Larger fuel pieces, given their larger mass, can absorb more heat, conducting it away from the 'fuel–air interface' and slowing the preheating phase of wood in the combustion zone.

Box 3.4. Idealised surface area-to-volume ratio

Consider three cylinders of wood all 1 m long, 0.1 cm, 1.0 cm and 10 cm, respectively, in diameter.

Diameter (cm)	0.1	1.0	10.0
Surface area (cm^2)	62.8	630	6440
Volume (cm^3)	0.79	79.0	7900
Surface area-to-volume Ratio	80	8.0	0.8

Note: Each order of magnitude increase in diameter produces an identical order of magnitude reduction in surface-area-to-volume ratio.

Other things being equal, the surface area-to-volume ratio relationship described above reflects a simple cylinder, but fuel particles come in all shapes and sizes. As the shape changes from a cylinder to something else (an extreme case would be a thin, flat fuel piece) the surface area-to-volume-ratio increases dramatically and the fuel burns all the more vigorously. The length of time a piece of fuel will burn is proportional to its thickness. So, for wood at 4–10% moisture, the residence time of flaming combustion – how long it will burn with flames – is around three times its thickness in centimetres; so a log 10 cm in diameter will flame for 30 minutes after ignition.

Smaller pieces ignite more easily because they need less heat to quickly increase their temperature. Larger pieces absorb more heat and conduct it away from the heat source, and so need more to get them to the ignition point. This is why we start fires with small bits of kindling and not whole large logs. Small fuels (referred to as fine fuels) are typically described as 5–6 mm (1/4 inch) in diameter or less – for example, small dead twigs or needles.

Temperature and energy

When people are asked to guess how hot a candle flame is, the answer is usually a few hundred degrees centigrade. In reality, flame temperatures burning in open air (i.e. not artificially fed extra oxygen as in an oxyacetylene torch) are somewhat hotter and, whether from a candle or a huge forest fire, are remarkably similar. Temperature can be estimated by the colour of the flame. A flame is hotter as it moves from a dull red (700 °C) to cherry red (900 °C) to bright yellow (1200 °C) to white (1300 °C) and finally blue-white (1400 °C/2550 °F). (The exception to this is the startlingly blue but cool flame found above glowing combustion where carbon monoxide is oxidised to carbon dioxide.) So the blue outer layer of a candle (where the fuel gas and oxygen mix) is at 1400 °C and the red portion in the middle near the wick is around 800 °C. Flames in forest fires, whether conifers in the north or eucalypts in the south, are normally of the same order, between 800–1200 °C (1500–2000 °F). Even the hottest forest fires are unlikely to go much above 1650 °C although temperatures up to around 2000 °C are possible under controlled laboratory conditions. (An oxyacetylene torch, which thoroughly mixes pure oxygen and acetylene gas under pressure, reaches almost 3500 °C.)

What is perhaps more important is not the absolute temperature but how much energy is involved in the fire. A forest fire clearly has much more energy than a candle flame. If you were crossing a road and a cyclist and a large truck were both coming at you at 10 miles per hour, you would probably prefer to be hit by the cyclist rather than the truck. Both are going at the same speed but the truck has a lot more energy behind it and dissipating that energy in stopping the vehicle will do you a lot more harm. In the same way the flame of a candle and a forest fire may have the same temperature but the forest fire contains a lot more energy.

(a)

(b)

(c)

Figure 3.3
A series of experimental fires were ignited in jack pine (*Pinus banksiana*) stands in Alberta, Canada in 1974 to test the link between fire behaviour and the Canadian Forest Fire Weather Index (FWI, described in Chapter 7). The resulting photographs show a series of fires at different fire intensities. (a) A creeping surface fire with flames only occasionally above a few centimetres high. The FWI was 8.5, the fire intensity 190 kW/m and the rate of spread of the front of the fire (the head fire) was 0.6 m/min. (b) A very intense surface fire, difficult to control even with heavy mechanical equipment with a number of individual trees igniting along their lengths (torching). FWI 24, fire intensity 1900 kW/m, rate of spread 3.4 m/min). (c) An active crown fire with 30-m high flames (10 m higher than the trees). FWI 34, fire intensity 7460 kW/m, rate of spread 6.1 m/min. From Alexander & De Groot (1988). Source: Canadian Forest Service (Natural Resources Canada), reproduced with permission.

So, how do we measure the rate of energy release of a fire, normally referred to as the 'fire intensity'? There are several ways of measuring this but the commonest is as 'fireline intensity', how much heat is being produced from a set length of burning fire front, expressed as kW/m (kilowatts per metre of fire front). (Americans use Btu/ft/s: 1 kW/m = 0.29 Btu/ft/s.) Mathematically, fireline intensity is the product of the weight of fuel consumed, the energy (calorific) content of the fuel (by weight) and the rate of fuel consumption.[1] In short, assuming a constant energy content, intensity is calculated by how much fuel is burning and how fast is it burning. As we will see later, fire intensity is one of the most important and widespread concepts in fire management. Not only is it used for suppression planning and action, it has been used extensively in developing an understanding of fire effects and fire regime (see Chapters 5, 6 and 7). Fire intensity has been described as the single most descriptive term for wildland fires (those that burn in the natural or semi-natural forest).

The lowest intensity of a fire still able to trickle along is around 10 kW/m and even modest fires (Fig. 3.3a) can develop several hundred kW/m. Fireline intensity can be fairly accurately predicted from the height of the flames (flame length). Thus a fire with the highest flames around 1.2 m high is burning at around 400 kW/m, just at the limit for control by hand tools. Flames up to 2.3 m tall will be at an

1. Fire intensity is calculated as $I = Hwr$ where I is the intensity in kilowatts per metre (kW/m); H is the energy content of the fuel (treated as a constant) expressed in kilojoules per kilogram (kJ/kg); w is the weight of fuel consumed over a set area, expressed as kilograms per square metre (kg/m²); and r is the rate of spread of the fire in metres per second (m/s). For further information see Alexander (1982).

intensity of 1700 kW/m, towards the upper limit of control by heavy mechanical equipment (Fig. 3.3b). Fires above 3500–4000 kW/m (Fig. 3.3c) are usually beyond the ability of suppression forces in direct attack (see Chapter 7). The upper limit of fire intensity in Australian eucalypt forest fires – probably the most intense fires in the world – is considered to be 100 000 kW/m, a figure 25 times higher than the maximum it is possible to control. Intensity is discussed further under the different types of fire below.

Anatomy of a fire

A fire and its parts have a descriptive terminology of their own (Fig. 3.4). Once a fire is burning, it will normally spread fastest in the direction of (with) the wind, slowest against the wind and in between speeds elsewhere. This produces a roughly elliptically shaped fire. The fastest spreading part going with the wind is referred to as the front or head, the slowest part burning against the wind is the back and the sides are the fire flanks. This leads to the expressions of a head fire or heading fire (burning with the wind) and a back fire or backing fire (burning against the wind). Each section of the fire has its own characteristics which will be described further in Chapter 6.

A fire starts at the ignition point, and it is usually a single point. Once established, it grows quickly (the build-up phase) to reach a speed of spread that becomes almost constant provided everything else remains constant. This is sometimes called a quasi-steady state phase; quasi or seemingly constant because other things are never constant and vary in time, such as weather (especially wind speed and direction) and space (for example, arrangement of fuel and its quality).

Key features of a fire burning steadily are the rate of spread of the head fire (usually timed over a set distance to get an average), intensity (as described above, the energy output above ground) and the severity (how much the fire burns beneath the litter into the peaty soil and dead wood).

To quickly characterise and describe fire behaviour, fires are often described in terms of their spread rate and intensity:

- **Smouldering:** a fire burning without flame and barely spreading.
- **Creeping:** a fire spreading slowly over the ground, generally with low flames.
- **Running:** a fire rapidly spreading with a well-defined head.
- **Candling:** a running fire with individual whole standing trees catching fire or flaring up.
- **Torching:** fire burning up into the canopy of a clump of trees.
- **Crowning:** a running fire with the tree crowns engaged in the fire.
- **Spotting:** a torching or crowning fire that is throwing fire brands ahead of the main fire front.

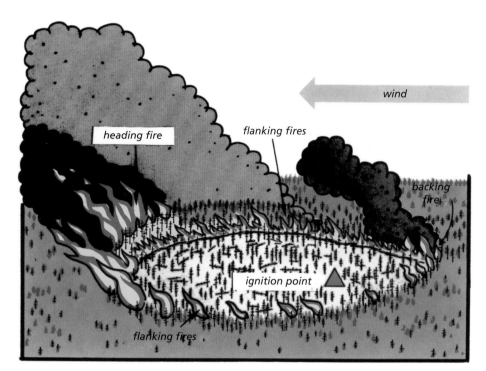

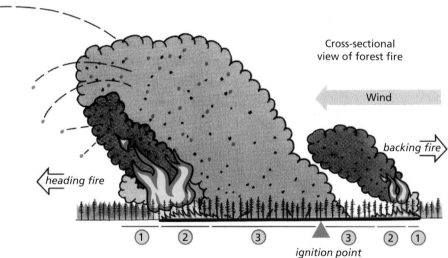

Figure 3.4
The top picture shows the main parts of a forest fire driven by the wind. The bottom picture below shows the same fire from ground level and shows burning embers being blown out in front of the heading fire which can be the start of new 'spot fires' described in this chapter. Points 1 are being pre-heated by the fire, points 2 are within the flaming fronts and at points 3 the flames are out, leaving glowing combustion behind.
From Cottrell (2004).

Types of fire

When it comes to understanding the effects of fire and the methods used to control them, energy output is important, but so too is the type of fire. Fires can be divided into three basic types, each named for the layer of the forest they burn through – ground fires, surface fires and crown fires.

Ground fire

These are slow-moving fires that creep beneath the surface litter through 'ground fuels' – roots, buried logs and the peaty layer of decomposing organic matter (called duff in North America) that builds up at the top of the soil especially in conifer forests (see Fig. 3.5). Ground fires can also burn in peat bogs when the fuel moisture is low enough and in coal seams. Because the volatile compounds and much of the cellulose are rotting away, the ash and lignin content is high and flammable gas production is low, so burning is primarily by glowing combustion.

Ignition is usually from a passing above-ground fire but a lightning strike onto a decomposing stump or similar can ignite the ground fuels. This small spark of fire slowly burns downwards and sideways by glowing combustion. The thin layer of glowing combustion creates a drying zone ahead with pyrolysis occurring at the junction of drying and glowing zones. The fire can burn, in effect, underground because glowing combustion produces carbon monoxide (see above), a reaction which can take place in an atmosphere with as little as 5% oxygen (normal atmosphere contains 21% oxygen) whereas flaming combustion, which produces carbon dioxide, needs at least 13–15% oxygen. Thus the progress of ground fires is governed by how quickly oxygen can diffuse to the glowing zone and by how much moisture there is in the organic matter. The moisture of extinction of smouldering combustion can be as high as 100–240% but it normally needs less than 40% moisture to continue.

Temperatures are lower than in flaming combustion: Zanon and colleagues (2008) investigated a fire in silty soils with 6–10% organic matter on the Island of São Miguel in the Azores, and found that in a week of burning, the fire reached a maximum temperature of 327 °C. Smouldering fires producing temperatures above 300 °C for 2–4 hours at any one point are not unusual. These relatively cool, low-intensity fires are generally very slow moving. Rates in the order of a fraction of a millimetre to a few centimetres per hour are normal but have been recorded in Australia as shooting along at 2 m per day.

As a ground fire burns, its downward path is eventually stopped by reaching the underlying mineral (rocky) soil or by increasing moisture with depth, although in well-drained or droughted areas this can be a matter of a metre or two. As it burns sideways, an overhang of unburnt material at the surface is often left (see Fig. 3.5a) because of heat loss to the air cooling it before it can burn. In fact, the fire may even leave a thin unburnt crust so that apart from the odd wisp of smoke, there may be no sign of fuel disappearing until someone walks over the area and plummets through the crust. Obviously this can be a dangerous situation; a person could unwittingly walk over the ground fire and fall a couple of metres into a pit of burning embers, dislodging the walls when they fall to bury themselves. These fires can burn beneath snow with the unburnt crust insulating the snow enough to stop it melting. The first sign of the fire only appears in spring!

Although ground fires are slow moving they can burn for months or years and are perhaps the most difficult fires to track and bring under control. Only copious amounts of water (most effectively, a rising winter water table) will put them out. They can also be extremely destructive. The large quantities of carbon monoxide

(a)

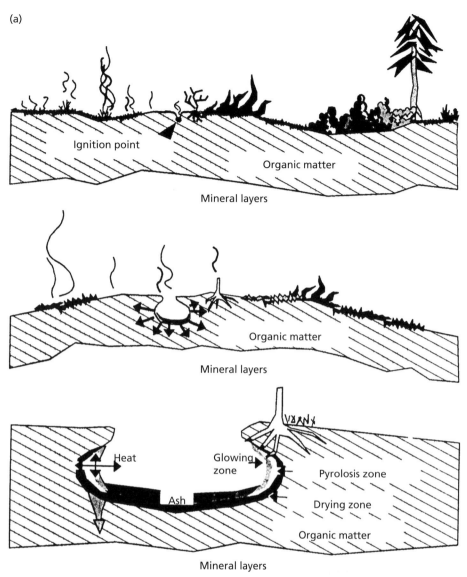

(b)

Figure 3.5
Diagram (a) shows how organic matter that has built up on the forest floor over the mineral soil burns by glowing combustion, starting from ignition to a well-developed fire with a distinctive steep edge surrounding the burnt out area (from Hungerford et al. 1995). An example of this steep edge is shown in the photo (b); the foot is on unburnt organic matter and pointing to the area where all organic matter has been burnt off of the mineral soil (this was taken a year or so after the fire and so the fire site is covered with debris that has fallen since the fire (taken in Jasper National Park, Canada). In the last two pictures, a ground fire on an island in the Slave River, northern Canada has burnt through the tree roots, causing the otherwise undamaged trees to fall over; this includes (c) a black spruce (*Picea mariana*) which shows the burnt-out roots, and (d) an overview of the felled forest consisting mostly of birch (*Betula papyrifera*). Hungerford et al. 1995 diagram © 1995 Tall Timbers Research, Inc. Reprinted by permission of the authors and Tall Timbers Research, Inc.

(c)

Figure 3.5 (*cont.*)

(d)

produced can kill animals, including humans. On a larger scale, they are also responsible for enormous quantities of carbon dioxide release into the atmosphere, an identified greenhouse gas, worsening climate change (see Chapter 8). The fires also relentlessly consume roots, seeds and soil organisms over large areas, causing the death and falling over of trees otherwise unaffected by fire (Fig. 3.5c, d).

Moreover, with an increase in oxygen supply from high winds these inconspicuous fires can give rise to a new flaming combustion fire above ground.

Even if the soil organic matter is too wet or sparse to burn, sizeable logs, buried or on the surface, can be burnt by ground fires until nothing is left but a line of ash. In the ongoing fires in the deep tropical peat of Indonesia, whole logs can burn out inside otherwise unburnt peat.

Surface fire

The majority of wildland fires start and finish as surface fires burning through the litter on or near the forest floor (freshly fallen leaves, needles, twigs, stems and bark – an especially important feature of eucalypts of the southern open forests of Australia) and downed woody material, including the 'slash' – tops and branches – left by forestry operations (Fig. 3.6). Lichens, grasses, herbs, shrubs and young trees are also all fodder for the fire, resulting in what can be a large fire with tall flames burning below, but not involving, the tree canopy. These fires can be intense and will generally burn dead branch wood up to 7.6 cm (3 inches) in diameter.

Crown fire

Crowning occurs when a fire jumps up into the crown or canopy of trees. Crown fires span a broad spectrum of fire behaviour, from single trees igniting as the surface fire spreads below, to large fires tearing through the tree tops with massive walls of flame from ground level to 30 or more metres above the forest canopy (Fig. 3.7). These are the fastest spreading and most dramatic of fires, usually short-lived but extremely dangerous. Speeds of more than 10 km/h (6 miles/h) with flames 20–30 m high are possible. Despite the high energy released by a crown fire it is mainly the living foliage and small-diameter dead wood that is involved in the fire. Consequently, they usually occur in forests with particularly flammable foliage such as the northern conifer forests and the eucalypt forests of Australia, but also the chaparral of south-west USA and other types of bush. However, even more important is how much moisture there is in the foliage.

Because of the typically high moisture of foliage in canopies, they tend to resist ignition unless certain conditions are met. First of all the canopy foliage needs to be dry enough (generally less than 100% moisture – see Box 3.3 for definitions) and enough heat energy needs to get up into the crown to ignite the foliage. This heat normally comes from a surface fire burning under the canopy (see Box 3.5). In the case of tree crowns, two heat transfer mechanisms are at work – radiation directly from the flames and upwards convection since the crowns are above the surface fire and in the flow of hot air and gases. Given that radiant and convective heat dissipate rapidly with distance from the source, it is logical to assume that the closer the crown is to the surface fire below, the more readily it will become involved.

(a)

(b)

Figure 3.6
Surface fires. Picture (a) is a very low-intensity fire trickling through the dead litter below an aspen stand (*Populus tremuloides*). (b) This shows a more intense fire burning through the litter and live vegetation near the ground. The flames are licking up the trunks of the red pine (*Pinus resinosa*) but the large gap between the surface fire and the canopy helps prevent this developing into a crown fire. Both pictures were taken in New Brunswick, Canada.

This therefore defines two major controlling factors for engaging the crowns in the fire – the intensity of the surface fire and the height of the base of the crown. Lower surface fire intensity narrows the vertical 'gap' it is able to jump to get into the crowns. If the surface fire intensity is higher, small trees with low crown bases ignite and 'candle' or a clump of trees will 'torch'. At still higher fire intensities, more and more of the trees become involved, until a full active crown fire is achieved.

(a)

(b)

Figure 3.7
Crown fires in dead and sick balsam fir (*Abies balsamea*) damaged by the spruce budworm in Ontario, Canada. (a) is a crown fire developing in this experimental plot that has been deliberately lit, and (b) shows that the fire involves a surface fire merging into a crown fire above. The shadow on the right is caused by the heavy smoke plume. Photographs courtesy of Brian Stocks, Canadian Forest Service.

Box 3.5. The three types of crown fire

Passive crown fire

Energy to keep the crown burning is provided by a surface fire burning under the tree canopy and slightly ahead to supply enough heat to the canopy. The crown fire is thus completely dependent on the heat transfer from below and the speed at which the surface fire moves controls the whole fire.

Active crown fire

With a little more energy available (less moisture in the fuel, warmer air temperatures, more fuel or a more intense surface fire) the crown and surface fires can spread simultaneously as a linked unit. The surface fire helps ignite the crown and the crown part of the fire helps preheat the surface fuels and so aids the spread of the surface fire.

Independent crown fire

With even better burning conditions, the crown fire may spread independently of a surface fire by the fire sweeping from crown to crown, each one producing enough heat to ignite the crown next door. This is helped by strong winds that bend over the flames, helping to preheat the next tree crown.

Since the conditions for each of these is progressively more exacting, it is not surprising that their occurrence as you go down the list is less common. Independent crown fires are extremely rare and many fire behaviour experts have doubted their existence. However, documented evidence from southern Alberta, Canada (the Panther fire) has led to the conclusion that these fires do exist. In this case it is postulated that warm 'chinook' winds in winter can melt snow and dry fuel in the open valley bottoms enough for crown fires to develop and then burn independently up slopes through the canopy above while the surface fuels are still too wet or snow-covered to burn. Nevertheless, an independent crown fire 'must be a rare and delicately balanced phenomenon, liable to collapse at any moment should one of the required conditions fail' (Van Wagner 1977).

As a rule of thumb, a gap of more than one and a half flame lengths will act as a vertical fire break. However, getting flames up into the canopy is helped by 'ladder' or 'bridge' fuels that give fire the means to bridge the gap between surface fuels and the canopy. Ladder fuels include such things as tall shrubs and saplings, flammable bark or lichens, and the retention of small branches and flaky or loose bark on trunks. That said, ladder fuels encourage torching or candling of individual canopies but will not by themselves start a crown fire unless other conditions are met. Fire tends not to spread in the canopy unless there is sufficient wind to deflect flames sideways or the ground is steep enough for flames to carry directly from one burning tree into the foliage of the next, and the canopies are close enough together for the fire to carry between them. So, in summary, a crown fire is more likely to develop where the foliage is flammable (rich in flammable components but most importantly low in moisture), the canopy is not too high above the ground, the surface fire is intense, the tree crowns are close to each other above ground and a strong wind is blowing. Since crown fires only occur when conditions are dry (producing long flames) and windy, they are always intense, fast moving and extremely dangerous. They are also very difficult to control by normal firefighting techniques and most often put out by a change in weather, fuel or topography.

Dynamics of extreme fires

Reports of extreme fires often involve words such as 'plumes' and 'fire whirls'. Just what are these and how do they influence fire behaviour? The word 'extreme' implies fast-moving fires, prolific crowning (see above) and generally unpredictable behaviour. Up until now we have been discussing the well-behaved (or at least predictable) fire, which is dependent on the fuel, surface weather and topography. These fires are known as wind-dominated fires, but there is another type of fire, however, the convection-dominated fire which has distinct atmospheric features such as a strong convection column and peculiar wind movements. It is usually difficult to predict how these fires will burn because of the enormous amount of energy they release and the fact that they often exercise some degree of influence on their environment and behave erratically.

Convection columns

A large fire produces enormous quantities of hot air that rises by convection. In the same way that the hot air rising above a candle collects together into a narrow stream, so the heat above a fire pulls itself together and rises as a column (Fig. 3.8). Usually the surrounding atmosphere resists this convective force due to high wind speeds aloft breaking up the column and the adiabatic lapse rate (which dictates that air cools as it rises), and the fire stays as a wind-driven event. But if enough

Figure 3.8
Convection column above Sioux Lookout Fire 48, Ontario, Canada. Photograph by Terry Popowich, used courtesy of the Ontario Ministry of Natural Resources. © 2003 Queens Printer Ontario.

energy is released, and the atmosphere is unstable (i.e. it does not follow the normal wind profile aloft and has an unstable temperature and pressure profile), a strong convection column can form, sometimes exceeding 15 km in height. If the vertical velocity of the air in the column is high enough, and the column is wide enough, it creates its own powerful surface winds pulled from all directions towards the column as air is dragged in to replace the huge mass of rising hot air. With a fire of this type, if you are standing nearby (within half a kilometre) facing the fire, you feel a strong wind at your back as the surface air all around the fire is drawn in to support the rapidly building convection column. You also hear a loud rushing sound, usually described as the sound of a locomotive. These surface winds can be remarkably strong and turbulent causing an increase in burning rates, which produces more heat and more convection in a positive feedback loop that can lead to a far more extreme fire than would otherwise be predicted. Crown fires are also virtually guaranteed.

Convection columns or 'plumes' can rise many kilometres into the atmosphere. Moderate winds near the ground will not affect the column and will flow around it almost as if it were solid. But stronger winds may lean-over or shear-off the top of the column. This is especially likely with height as the winds get stronger aloft. Convection-driven fires are unpredictable and are uncontrollable. But this is not their only cause for concern; convection columns also help produce spot fires and fire whirls (see below). Eventually the convection column weakens as the fire runs out of fuel or weather conditions change. At this point, the column can fade away but if the weakening is very rapid it can dramatically collapse with very strong downdrafts which can carry burning embers several kilometres in all directions.

Spot fires

Spotting occurs when burning embers (firebrands) are carried some distance away from the fire front where they ignite surface fuels, resulting in new fires – spot fires or 'jump fires'. Three basic elements are required for spot fires. First, there must be some fuel that can be carried aloft while on fire, and carried away from the main fire. The key to a successful spot fire is the ability of a burning ember to land with enough heat left to kindle a new fire. Fuel moisture is thus critical for spot fires. Too much moisture, and a burning ember will not be able to start a fire where it lands; too little, and the burning ember will burn too quickly and go out before it lands. In eucalypt forests a moisture level of 4% is optimum for spotting. The fuel must also physically be in a position from where it can be carried away. Fuels lying on the ground require a lot of turbulent wind to lift them up, such as fire whirls (see below) or very strong winds. Burning pieces already above the ground (such as leaves, cones, twigs or bits of bark) are more easily caught by rising hot air and carried off and are the source of most spot fires.

The second element needed for spot fires to occur is sufficient convective energy to carry the firebrands aloft (Fig. 3.9). Typically the higher the embers are carried, the farther away from the fire front they will end up. Strong horizontal winds are capable of starting spot fires but these tend to be fairly close to the fire front, since the burning embers hit the ground pretty quickly in even the strongest winds. Convection columns, however, because they can lift burning fragments high into the atmosphere, can result in spot fires a considerable distance away. This is especially true if there is a wind blowing to carry the embers horizontally once they are released from the convection column or the wind is strong enough to bend or break the convection column as in Fig. 3.9b. The third and final element needed is that the ground where the firebrand lands must be receptive to rapid ignition: dry fine fuels perhaps with some wind to fan the flames.

The maximum spotting distance from a blaze depends on a number of factors: how high an ember is carried by the fire, horizontal wind speed, how fast the ember falls vertically, the initial size of the ember, its moisture and how rapidly it is burning. Short-range spotting is common in intensive fires with some crown involvement, long-range spotting is more common in large convection-driven fires. Short-range spot fires are less dangerous because they are likely to be soon overrun by the main fire and go largely unnoticed.

Australian eucalypts have an impressive record for encouraging spot fires. Mountain ash (*Eucalyptus regnans*) produces up to 1.5 tonnes of bark per year, often borne in large clumps more than 70 m above the ground. The messmate stringybark (*Eucalyptus obliqua*) of south-eastern Australia has a reputation for extensive spotting up to

(a)

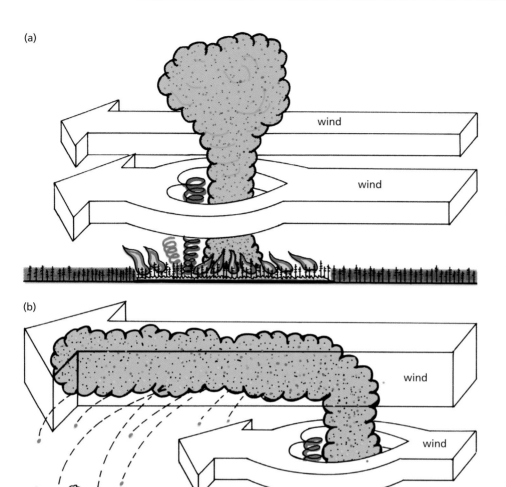

(b)

Figure 3.9
When a convection column above a fire is very strong, surface winds are forced around the outside of the column as shown in (a). This helps create eddies around the edge of the column (shown in blue) which can be responsible for fire whirls (see the text for a description of these). The fire whirls and the column can carry burning embers high into the air which, if the conditions are right, can drop to the ground and start new fires – spot fires – some way from the original fire. If the wind is strong enough to break or bend the convection column (as in b) then the chance of spot fires some distance away from the original fire is greatly increased. Burning embers can land at the rate of many thousands per hectare. From Cottrell (2004).

5 km away. Burning bark fragments from a fire about a kilometre away are thought to have been responsible for one fire which destroyed 100 houses during the 1994 bushfires around Sydney. Research has shown that in intense fires it is the larger, slow-burning pieces of bark that are most likely to start spot fires since they can typically burn for up to 6 minutes, and so can be carried high up in the intense convection column and still be burning when they land. In less intense fire conditions, it is the smaller fragments that are more dangerous; smaller pieces are carried higher than bigger pieces (but not so high that they burn up before landing) and so will be carried further away, starting spot fires further away than bigger fragments.

The candlebark eucalypts of Australia (such as the manna gum, *Eucalyptus viminalis*) are aptly named and notorious for causing spot fires. The bark dries, curls into hollow candles and hangs in long ribbons in the canopy of the tree. These are readily ignited and in hot fires are torn off the tree and can be carried 3–4 km into the air, still burning, easily producing a spot fire 30 km or more away.

Spot fires can be particularly dangerous since they can start behind firefighters or those fleeing the fire, trapping them in between fire fronts or in difficult and dangerous conditions. If the original fire has a large convection column, this is made potentially worse because the new fire is drawn to its creator with great speed and intensity (due to the induced surface winds), reinforcing the extreme fire. Spot fires are blamed for many firefighter disasters, including the tragic Mann Gulch Fire in Montana in 1949, where 13 smokejumpers died (the full story of the Mann Gulch Fire can be read in Norman MacLean's 1992 book *Young Men and Fire*).

Fire whirls, horizontal roll vortices and fire storms

If heated air rising rapidly from a fire is given an initial start-up spin, it can rapidly develop into a fire whirl, a swiftly spinning column of rising hot air (Fig. 3.10a). Fire whirls (or vertical vortices) range in size from less than 30 cm in diameter (very common) to more than 150 m (very rare) with the intensity of a small tornado. They are thus capable of picking up large quantities of burning material (whole branches and bushes, not just small twigs and cones), and are consequently potent forces in initiating spot fires.

What causes the initial spinning of the air to start a fire whirl? It can be the friction from a rough surface, such as a patch of shrubs, slewing one side of a gust of wind around to start it spinning, like a car clipping a tree and spinning out of control. Fire whirls also commonly start just on the lee side of a ridge where convection currents meet winds coming over the ridge. Helicopters may also be linked to the starting of fire whirls. Once started the warm air rising pulls the whirl into a tight column, increasing its spin speed like an ice skater pulling in their arms to spin faster.

There are two broad classes of fire whirl. The first and commonest type is relatively small in diameter, ranging in height from a metre or so to around 25 m, and is short-lived with a lifespan of less than a minute. These are the common occurrences resulting from extremely rapidly rising air pockets. The second type of fire whirl is much bigger and is found around large convection-column fires. This category has two subgroups. The first subgroup forms on the lee side of the column – the result of wind eddies around the column and pockets of rising air, similar to water swirling around a rock in the stream. The second subgroup is when a large convection column (more than 100 m wide) begins to rotate in its entirety; these can be very powerful and dangerous (like a very large tornado). Fire whirls can be very destructive. As an example, in September 1923 an earthquake hit Tokyo leading to several large fires sweeping through the city; 40 000 people took refuge in a large open area in the city but were hit by several fire whirls started by winds eddying around buildings, killing 38 000 people within 15 minutes (see Kuwana *et al.* 2007).

Hot air from a forest fire can spin horizontally as well as vertically. When it does so in conjunction with a crown fire it can create a 'horizontal roll vortex'. These, unlike their tornado-like cousins, form on wind-driven and extremely rapidly moving fires

(McRae & Flannigan 1990). Cool winds interacting with hot air from the fire create a mass of air that rolls as it moves forward through the forest (Fig. 3.10b). It produces a strong downdraft in a line that leaves unburnt 'tree crown streets' parallel to each other where the understorey shrubs and canopy are left unburnt. Charring on the trees suggests that air moves down from above, through the trees and then out sideways. Tree crown streets were recorded in the Mack Lake Fire in Michigan by Haines (1982) as being 10–200 m wide, 45–850 m apart and 3–6 km long.

If all the factors of an intense fire are put together, they produce a picture of a very dangerous situation. An intense fire can produce a large and strong convection column which in turn gives rise to tornado-like fire whirls and aggressive spotting. The new spot fires are drawn towards the huge convection column to create intense fires with hot plumes of air rising from them, which may quickly merge. The turbulence caused by wind and rapidly rising air in turn produces violent and gusty winds near the ground causing unpredictable bursts of intense fire in different directions. To this can be added the influences of topography on fire behaviour (see Chapter 4), which includes rapid fire spread up slopes, and swirling winds through valleys and over ridges. This is in short, a firestorm: uncontrollable, unpredictable and very dangerous.

Scales of fire impact: smoke

Finally in this chapter, we finish with a thought on the different scales of impact that fire has on the environment, taking the example of smoke. At the smallest scale smoke movement from a small fire varies in amount and direction of travel. This has important influences on our enjoyment of a campfire with stinging eyes and inhalation of smoke. More than 100 compounds have been identified in wood smoke including a variety of aldehydes (such as formaldehyde and acrolein) and phenolics, both of which are noted lung irritants, and polynuclear aromatic hydrocarbons, some of which are known to be carcinogenic, so breathing in smoke is generally to be avoided (see Goh *et al.* 1999, Radojevic 2003 for more details). These compounds are produced mainly by glowing and low-intensity flaming fires. Factors that affect smoke production at this scale include what species are burning, how wet the fuel is, whether it is burning by glowing or flaming combustion and the vagaries of wind gusts. Where the smoke lands at this small scale (either directly or dissolved in rain) can also have important implications for the regeneration of plants from seeds that need smoke to germinate (see Chapter 5).

At the landscape scale (Fig. 3.11a) the short-term smoke plume from a fire has equally important features, especially when it is considered that large fires have been recorded as producing more than 0.6 tonnes of particulate matter per second into the atmosphere. Smoke can degrade air quality, impair visibility and worsen regional haze. Smoke also reduces the amount of the sun's energy reaching the ground and hence affects wind and weather patterns; temperatures at the soil

Figure 3.10 (opposite page)
(a) A fire whirl during an evening fire at Eastern Virginia Rivers National Wildlife Refuge Complex. Photograph by Gary Kemp, US Fish & Wildlife Service. (b) Horizontal roll vortices. The cross-section shows hot air rising from the fire on the right circulates round and drops as it cools, bringing in cool ambient air from outside the fire to create a strong downdraft. The line of trees directly under the downdraft escape burning and remain as an unburnt 'crown street'. From Haines (1982).

(a)

2/19/2002 18:22

(b)

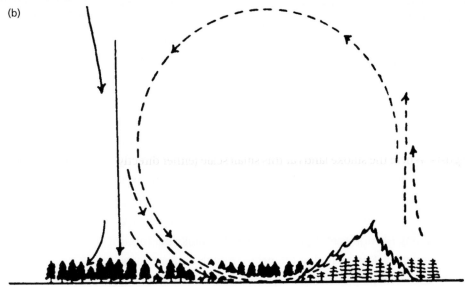

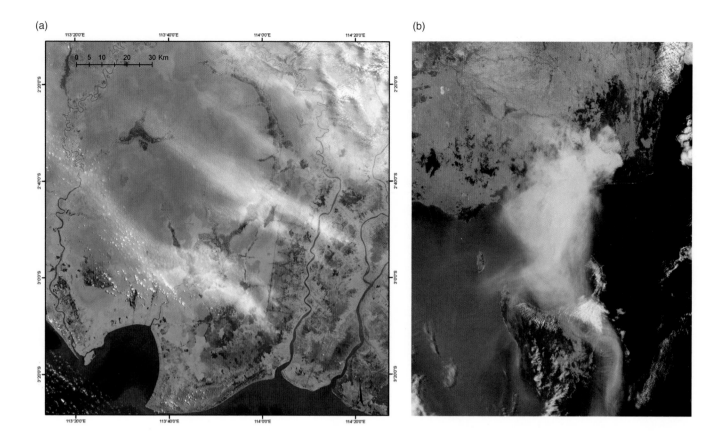

(a)

(b)

surface have been measured at 5–15 °C lower than normal after large fires. Smoke also contains chemicals remobilised by the fire, such as lead and other heavy metals; the hotter the fire the higher the proportion of elements that appear in smoke and will be redeposited elsewhere. Small particles less than 10 μm (0.01 mm) make up less than 10% of the total weight of smoke particles but are the most critical for health (Karthikeyan *et al.* 2006). Fine particulate matter (particles less than 2.5 μm in diameter – 0.0025 mm, often referred to as PM2.5) are the 'respirable fraction' that are drawn deep into the lungs when we breathe, with potentially severe long-term health consequences (Sastry 2002), and these are the particles which will travel furthest once the larger, heavier particles have dropped or been washed down to the ground, and remain in the air the longest (weeks rather than days, and into years in the high atmosphere). This is also the size of particle that has the most effect on visibility in the atmosphere, and can be so dense that large fires in countries such as Brazil and Indonesia (see See *et al.* 2007) have resulted in airports being closed. Factors affecting smoke production at this level include what sort of forest or bush is burning (shrubby areas produce the most smoke per tonne of fuel consumed) and its intensity. Fires of higher intensity, i.e. with longer flames, produce proportionately larger smoke particles (i.e. less PM2.5) than low-intensity or smouldering fires.

Figure 3.11
Smoke plumes as seen by satellite. (a) Image from the UK-DMC satellite of tropical peatland fires in Central Kalimantan, Borneo (Indonesia) on 11 September 2005 (ID scene DU000607p_L1R). UK-DMC image © 2009 SSTL, supplied by DMCii. (b) The picture on the right was taken on 4 February 2003 and shows fires in south-east Australia. The plume was 300 km (186 miles) wide in places and more than 877 km (545 miles) long, stretching across Tasmania to the south dots, had been burning out of control in the region for several weeks. Images courtesy of The Visible Earth, NASA (visibleearth.nasa.gov/).

At a global scale, smoke plumes can cover very large areas, crossing international boundaries. These can affect large-scale weather patterns by reducing the amount of the sun's energy reaching the ground. The smoke also contains greenhouse gases implicated in global climate change; in fact global biomass burning is estimated to contribute 2–3.3 billion tonnes (= Giga tonnes, Gt) of carbon to the atmosphere each year. The *Astronaut Training Manual (Atmospheric Phenomena)* prepared by the Johnson Space Center (see Lyons *et al.* 1998) notes that smoke palls larger than 300 000 km^2 were unusual during the late 1960s but by 1991 were commonly 10 times bigger than this (more than 3.3 million km^2); slightly less than half the area of Australia. Figure 3.11b shows that smoke plumes can be very large indeed, potentially encircling the globe. Of course, from the ground, much of this plume would be almost invisible but it will still be interacting with ground weather patterns and atmospheric chemicals producing ozone-rich smog. These plumes can travel a thousand kilometres in a few days and persist for weeks.

4 | Fire in the wild landscape

The previous chapter explains how a fire burns and what types of fire are possible. But once out in the real world there are many influences that determine whether a fire starts and how it subsequently behaves. First of all, there has to be a source of ignition – the spark or heat source intense enough to cause combustion – and this has to coincide with some sort of flammable material in a dry enough state to burn. Once the fire has started, its behaviour in terms of how fast and far it will spread and how much energy it produces is primarily controlled by the weather, the shape of the terrain (topography) and the amount and state of the fuel. These three factors can be fitted together as a fire behaviour triangle (see Fig. 4.5); change one side of the triangle and the fire regime will also change, often dramatically. Which one of these factors is most important in shaping the fire of an area will vary depending on the region, the ecosystem type and historical events such as previous fires.

The fire regime of an area can be defined as the characteristic frequency, extent, intensity and seasonality of fires within an area. You should also be aware that the fire itself can influence the weather and fuel sides of the triangle and hence the fire regime – it is a two-way process. For example, a vigorous fire can make its own wind and clouds; these clouds can in turn produce lightning to start more fires, and rain to put them out.

Causes of wildfire – how do they start?

Lightning

The perhaps surprising answer is that the biggest single natural cause of fire ignition is lightning. Surprising, in that lightning is not that common for most of us and it is usually associated with rain. Nevertheless, lightning is a powerful force, as reflected in mythology where the Greeks feared the lightning flung by Zeus, the Vikings thought lightning was produced by Thor as his hammer struck an anvil, and aboriginal tribes in North America believed that lightning was from the flashing feathers of a mystical bird. Despite appearances, lightning is remarkably frequent, and at any one time there are likely to be nearly 2000 thunderstorms occurring throughout the world and lightning hits the ground somewhere in the world at least 8 million times a day, an average of 100 lightning strikes per second

(a)

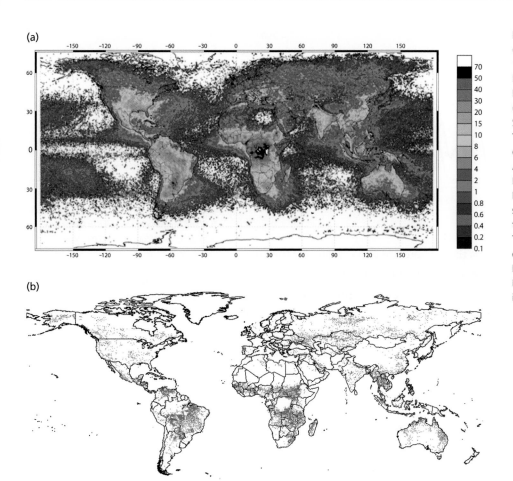

(b)

Figure 4.1
(a) Global distribution of lightning between April 1995 and February 2003, showing the number of lightning flashes per square kilometre per year. Image courtesy of The Visible Earth, NASA (visibleearth.nasa.gov/), produced by the National Space Science and Technology Center Lightning Team. (b) Global distribution of fire in 2005. The data are collected at night by the ERS-2 satellite's Along Track Scanning Radiometer (ATSR-2) and Envisat's Advanced Along Track Scanning Radiometer (AATSR). These twin radiometer sensors measure thermal infrared radiation, taking the temperature of earth's land surface. Temperatures exceeding 38.85 °C at night are classed as burning fires. Data are from the World Fire Atlas from the Data User Element of the European Space Agency (http://dup.esrin.esa.int/ionia/wfa).

(Taylor 1971). These strikes are not evenly spread around the globe, as Fig. 4.1a shows; lightning is most common in tropical continental areas of South-East Asia, northern Australia, Africa, Central and South America. Fortunately for us, these are primarily rainforest areas where the wet climate usually keeps the fuel too wet to burn, so the incidence of lightning does not entirely match the pattern of forest fires (compare Fig. 4.1a with 4.1b).

Even in areas with drier climates where fires are common, such as the dry shrublands of the Mediterranean and Australia, and the coniferous forests of North America and Siberia, very few lightning strikes actually start a fire; perhaps only one in a hundred to less than one in a thousand, although this does vary tremendously; Wierzchowski *et al.* (2002) show that in Alberta, Canada it takes 1400 strikes to start a fire, but across the border in British Columbia, it takes fewer than 50 strikes. Beth Hall (2007) worked out that between 1990 and 1998 over 17 000 naturally ignited wildfires occurred in Arizona and New Mexico on US federal land during the fire season of April to October. The lightning strikes associated with these fires accounted for less than 0.35% (1 in 285) of all recorded cloud-to-ground

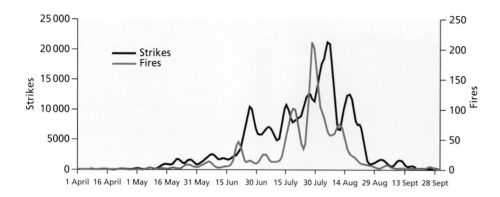

Figure 4.2
Since lightning is a prime cause of forest fires, it is perhaps not surprising that the numbers of lightning strikes and fire tend to coincide. The graph shows number of lightning strikes and fires between 1989–1994 summarised by date during the fire season for an area of southern British Columbia and Alberta centred on the Rocky Mountains. Although not completely in phase the lightning and fire occurrence both peak at the beginning of August. The slight shift in the lines is due to there being a number of other factors that dictate when fires burn (such as the dryness of the fuel and humidity of the air). © International Association of Wildland Fire 2002. Reproduced with permission from the *International Journal of Wildland Fire*, **11**(1): 41–51 (Wierzchowski, J. Heathcott, M. & Flannigan, M.D.), http://www.publish.csiro.au/nid/114/paper/WF01048.htm. Published by CSIRO Publishing, Melbourne Australia.

lightning strikes that occurred during this time. There are two reasons for this low success rate. First, not all lightning strikes that hit the ground are of the same quality and capable of starting a fire (see below for an explanation). Second, the fuels must be fine and dry enough to sustain combustion, and lightning storms are often associated with heavy rain. For a lightning strike to be successful in starting a fire, it must either: (1) hit dry ground away from the rain produced by the storm (which is often quite localised); or (2) ignite a rotting log, wood in a lightning-struck tree, or protected patch of dry organic matter on the ground. Rain-protected fuel can smoulder for many weeks or even months until the surrounding fuels dry out enough for flames to break out and spread (referred to as a holdover fire). In Finland, observations by the firefighting services between 1998 and 2002 showed that 89% of the recorded lightning-ignited forest fires started with a strike to a tree (the other 11% were on human-made structures or the ground). Of those tree strikes, 97% were on live trees, with the fires starting in 80% of cases on the ground at the base of the tree (Larjavaara *et al.* 2007).

Although lightning may seem a most improbable way of starting forest fires, empirical evidence shows that it works (Fig. 4.2). In the next chapter we will see that many plant species are adapted to periodic natural fire – the source of which is lightning. Details of how it works are given below but two examples show the importance of lightning. A lightning storm on 30 August 1987 in south-west Oregon and northern California was responsible for 1600 fires. On the other side of the coin, the shrublands of Chile (the matorral) are unique compared with those of the rest of the world (such as the chaparral of California, maquis of Mediterranean Europe, the fynbos of South Africa) in having very few natural fires – primarily due to the rarity of lightning-producing thunderstorms.

How lightning works Large thunderclouds build electrical charges – normally negative at the bottom of the cloud and positive at the top. The majority of lightning flashes are from cloud to cloud but a significant, and from our point of view, important, proportion go from cloud to ground (Fig. 4.3). Like electrical charges, there are two broad types of lightning discharges – negative and positive. A normal lightning flash has a negative charge and involves the bottom of the cloud. Streams of

Figure 4.3
Multiple cloud-to-ground and cloud-to-cloud lightning strikes during a night-time thunderstorm. Image courtesy of the National Oceanic and Atmospheric Administration Photo and Central Libraries, USA and OAR/ERL/National Severe Storms Laboratory (NSSL).

negatively charged ions protrude from the bottom of a thundercloud like static hair. A lightning strike starts from the ground with a sudden bridging of the gap rather like an arc forming between two electrodes as they are brought closer together, resulting in a strong negative charge (up to 100 000 amps) returning to hit the ground. The passage of electricity in this 'return stroke' has so much energy that the air is superheated – which gives the light and the thunder as the hot air expands outwards. During the first few microseconds of a return stroke, the 1–2 cm-wide core is heated to a maximum of about 30 000 °C – five times hotter than the surface of the sun – with a peak power of upwards of 10^{12} watts (this is 1 with 12 zeros after it, a million, million watts). The energy contained in a single strike of this magnitude is enough to boil half a million litres of water in a domestic kettle!

Although hot and full of energy, this main strike is so short that it is inadequate to start a fire. However, the majority of negative flashes are multistroke, composed of usually 4 but up to 30–40 secondary strikes along the same path of the main stroke at intervals of around 30 milliseconds; but they are all very similar to the main strike, and even cumulatively they do not start fires. However, some lightning flashes (usually called 'hybrid flashes') contain a few strikes that have a small but sustained current (a 'continuing current') of 100 amps or so which flows into the ground for up to half a second. This may not appear very long or powerful but the temperature is still between 6000 and 12 000 °C, and lasts long enough to be capable of igniting dry fuel. About 1 in 3–5 (20%–30%) of negative flashes are hybrid flashes capable of starting fires.

As important, but in a different way, are 'positive' flashes, which tend to occur particularly towards the end of a thunderstorm's life. These start not in the negatively charged bottom of the cloud but in the positively charged top of the cloud. Lightning from this area effectively carries a positive charge to the ground, usually made up of just one stroke, often with a high current. Most importantly, about 95% of positive flashes have the continuous current responsible for igniting fuel (compared with 20–30% of negative flashes). Consequently, almost all positive flashes have the capability of starting fires. Positive lightning also usually strikes at an angle away from the cloud, hitting the ground as far as 8–30 km (5–20 miles) away (hence the expression 'a bolt from the blue [sky]'!). This increases the chance of the strikes hitting dry burnable material not wetted by the storm. It would be wrong, however, to assume that it is only positive lightning that causes fires, and a number of studies have found no link between the number of fires and the rate of positive strikes (e.g. Flannigan & Wotton 1991). Bear in mind that positive flashes normally constitute only about 5–10% of all cloud-to-ground flashes and so negative lightning may ignite as many fires as positive lightning. In Finland, a study of 522 fires between 1998 and 2002 found that positive and negative strikes were equally probable to ignite fires: 11.1% of all strikes were positive and these caused 11.8% of fires (Larjavaara *et al.* 2007).

Having said this, under certain conditions, positive lightning takes on a much more meaningful role as a potent ecological force. Fires can produce their own thunderstorms; the hot air rising above a forest fire can result in cumulus clouds forming ('pyrocumulus') which can become big enough to develop into thunderheads. It also happens that thunderclouds forming in smoke-contaminated air produce vast amounts of positive lightning with increased potential for starting new fires (Lyons *et al.* 1998). An example illustrating this is shown in Box 4.1.

Other lightning effects Lightning can influence fires by creating dead wood and litter for future fires. Lightning has been seen to kill or injure groups of trees, sometimes up to hundreds of individuals. This has been seldom reported from the USA but has been commonly observed in parts of Europe, Australia, New Guinea, Malaysia and Central America. These 'lightning gaps' in the forest can be a source of easily ignited fuel for subsequent fires.

A single lightning-struck tree can also be a breeding ground for beetles, leading to mass attacks on surrounding trees, in turn leading to the build-up of burnable fuel with predictable results.

Volcanism

In areas of the globe where volcanoes are active, flowing lava can ignite fuel, be it forest, grass or otherwise. Molten rock flowing as lava has a temperature typically of 600–1100 °C backed by a lot of energy (see Chapter 3). Any vegetation or building that it meets, regardless of the moisture content or size, will inevitably be set on fire. But this is, of course, local to the lava flow. Ash-falls can also be hot enough to start fires, even after their journey through the air, but this is still limited

Box 4.1. Example of a fire breeding more fire by smoke producing increased positive lightning

The huge smoke plume above North America shows this working on a large geographical scale. The plume moved from southern Mexico and Central America across the USA and up into southern Canada in 1998. The thunderstorms forming along the plume did not produce an abnormal amount of lightning strikes but up to 85% of cloud-to-ground flashes were positive compared with 5–10% in cleaner air to the north. This resulted in nearly half a million positive flashes in the southern plains during a 2-month period, each potentially capable of starting a fire. The same phenomenon was reported during the severe Indonesian fires of 1997 and has been seen elsewhere in the world. It is food for thought that smoke is not the only cause: increased amounts of positive lightning are also being seen downwind of many Midwestern US urban areas, presumably due to atmospheric contaminants (smog).

The map was generated from Geostationary Operational Environmental Satellite (GOES) data in visible light taken at 2245 UTC, 15 May 1998. Courtesy of the Space Science and Engineering Center, University of Wisconsin, Madison.

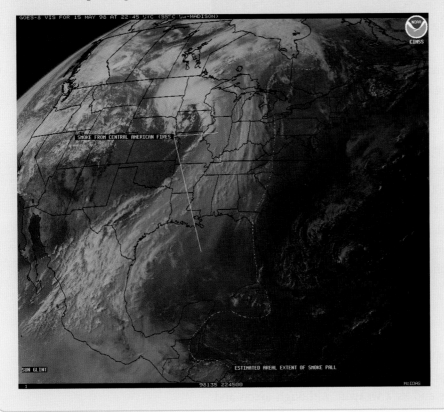

to fires just a few km from the source at best. Most records of fires started by volcanism come from Hawaii (Ainsworth & Kauffman 2009), East Asia (Japan, Indonesia) and East Africa (plus less often in the west).

Another example can be seen from an eruption of Mount St Helens in Washington State, USA 2000 years ago. Lava flowed down the mountain's south slope and

Figure 4.4
Lava flowing from an eruption of Mount St Helens, Washington State 2000 years ago flowed around pine trees and set firm enough to retain the shape of the tree as the trees burnt from the intense heat. Lava encased the smouldering tree leaving a raised lip around a pillar of charcoal which gradually eroded away. The resulting holes show the shape of the tree including the radiating root channels. Some of the holes bear the pattern of the bark preserved in the lava. Even after all this time, charcoal can still be found in some of the root channels.

around the base of trees on the lower flanks of the mountain and set them on fire. We know this because the lava was cool enough to congeal as the trees burnt, leaving tree-shaped holes, complete with root channels and the imprint of bark, in the lava (Fig. 4.4).

The relationship between volcanic activity and fires is not always straightforward, as illustrated by the more recent explosive eruption of Mount St Helens on 18 May 1980. Trees were blown down over 59 600 ha (230 square miles) by an ash- and rock-laden blast cloud that raced across the landscape, and which effectively smothered any fire it might have started. Dozens of fires *were* started but these were caused by lightning produced from the boiling, vertical ash cloud (which reached 25 km (16 miles) into the atmosphere).

Human fire starters

As detailed in Chapter 2, aboriginal peoples have been setting fires for tens of millennia as a hunting tool, to encourage beneficial plants for grazing and food, to allow easier travel, spying of enemies and as a weapon. Although arson and carelessness are blamed for an increasing amount of forest fire, the evidence suggests that aboriginal peoples of North America and Australia, in particular, were far better at burning the landscape than we are!

Such deliberate management of the landscape is still being carried on in developing countries and increasingly so in developed countries since the 1980s, when over much of the world the natural role of fire in many ecosystems was rediscovered. Fire has a useful role to play in maintaining conservation areas and in reducing fuel loads as a preventative measure against hostile wildfires. Now we are re-learning what our forebears knew very well.

Unfortunately, while we are re-learning the controlled application of fire, the majority of human-started fires are caused by accidents or carelessness. It would be difficult to list all possible inventive ways that we ignite and let fires escape, but high up the list come items such as discarded cigarettes and neglected campfires. Two of the largest fires of 1988 in Yellowstone National Park (the North Fork and Hellroaring fires) began with a tossed cigarette and an untended campfire. Obvious other causes include chainsaws sparking against rocks and sparks from other equipment such as shorting power lines and trains. Less obvious, perhaps, are such things as fireworks and parking vehicles in long grass which is ignited by contact with hot exhaust systems and catalytic converters.

As well as accidents there is intentional arson, whether deliberately malicious or a bit of fun that gets out of hand. Arson can be used to seek financial gain whether it is bogus insurance claims or seeking employment as a firefighter. Box 4.2 gives a few of the almost endless examples. According to the National Interagency Fire Center, 43% of the area burnt by fire in the USA in 2000 was due to human-caused fires.

Accidental fires and arson are inevitably more numerous near centres of high population or recreational use. Malcolm Gill and his colleagues (1987) found that in the eucalypt forests of Victoria, Australia there were more fires on weekends and public holidays than would be predicted from the weather.

Other sources

While the above are the main causes of fires, there are others. Sparks from rock collisions in landslides have been reputed to start fires – from North America to South Africa. As an example, a landslip during the night in Owens Valley, California, as part of the famous earthquake of 1872, produced a shower of sparks impressive enough to convince locals that an eruption was happening.

While on a geological theme, coal seams that come to the surface can catch fire with long-term consequences. For example, coal seams have been smouldering in Indonesia for the last 13 000–15 000 years, repeatedly kindling new vegetation fires when fuel conditions permit. And on 7 June 2002, a fire near Glenwood Springs, Colorado which eventually covered 4300 ha (10 600 acres) and burnt 28 homes worth $4.5 million, was started when a long-smouldering underground coal fire burnt through to the surface and raced through brush and trees.

As pointed out in Chapter 2, meteor impacts can start fires. A meteor that exploded 30 km above the ground north of Irktusk in Siberia on 25 September 2002 released the energy equivalent to 200 tonnes of TNT and resulted in a 100-km^2 area of burnt and toppled trees.

Power lines that touch each other can produce enough sparks to start fires. A study in Ku-ring-gai Chase National Park near Sydney, Australia, found that while 'clashing of powerlines' was a minor cause of fire (less than 1% of the number of fires) it was a major contributor to the total area burnt (36%). The average area of fires started by powerlines (440 ha) was more than four times greater than that from arson-caused fires (Conroy 1996). Powerlines hitting together can start fires by producing showers of hot metal and insulation falling to the ground.

Spontaneous combustion is also possible in large piles of moist grass, logging debris, wood chips and sawdust. Microbial activity, aided by respiration of live cells in the pile (especially in leaves), can raise the temperature to about 70 °C. Chemical oxidation then takes over and, if the pile is dense enough to stop the heat escaping, charring and smouldering combustion will start inside at 280 °C which can break out as flaming combustion. One of us (PAT) can remember putting his hand into a pile of chipped branches and leaves just a few days after it had been piled up and had started to steam. The hand didn't stay there long! Spontaneous combustion is a common threat to agricultural hay producers; hay must be well dried before it goes into the barn, or a fire is likely. Generally hay needs to be below 14% moisture for large bales and between 18–20% for small rectangular bales to be safe (Mickan 2006). A four-week period of 2007 saw 200 hay stack and shed fires across New South Wales, Australia started by either spontaneous combustion or by storage-related activities (Scott Keelan, personal communication).

What starts most fires?

This is a complex question that includes elements of the climate (conducive weather), population density and lightning incidence. A look around the world (see Box 4.3) shows that in countries that have areas of low population density, such as Australia, Canada and even parts of the USA, lightning is still the major

Box 4.3. Fire incidence by source on 'forest, other wooded land and other land' in different parts of the world. The data are taken from the *Global Forest Fire Assessment 1990–2000*, a report produced by the FAO (Food and Agriculture Organization of the United Nations) – Anon (2001). '?' indicates an unknown value.

Country	Causes of fire (by % of numbers of fires)		
	Natural	Human	Unknown
Africa			
Ethiopia	0	100	0
Kenya	0	60	40
Mozambique	~5	~90	~5
Asia			
India	?	99	?
Laos	?	90	? Slash and burn cultivation
Japan	?	99	?
South Korea	0	80	20
Oceania			
Australia	26	51	24 (Of the human-caused fires, 25% were deliberate and 26% agricultural sources and campfires)
Fiji	1	99	0
New Zealand	0.2	98	2
Europe			
Portugal	3	77	20 (43% negligence, 34% arson)
Spain	<10	66	>24
Italy	1	63	36
Estonia	1	81	18
Cyprus	17	46	37
Turkey	3	48	49
Germany	2	59	39
Russia	16	81	3
Finland	10	61	29
Americas			
Canada	35	?	?
USA	12	88	0
Mexico	3	94	3
Cuba	11	47	42
Bolivia	0.1	99.9	0
Argentina	11	61	28

cause of natural fires, especially away from humans. Overall in Canada 35% of fires are caused by lightning and in Australia 26%. Yet in remote parts of Australia, lightning has been reported as causing 80% of all fires, compared with just 5% in the densely populated Australian Capital Territory (ACT).

Box 4.4. Patterns of lightning and human-caused fires in north-west Spain

The map (b) is a 3D representation of the number of fire ignition points (ip) per square kilometre over a 19-year period from 1983 to 2001 in the area shown in (a). Each ignition point is classified as to whether it was human or lightning caused and the resulting map blends the colour in proportion to the two types. It is clear that the lightning fires tend to be clustered, particularly in the south-east portion of the Iberian Mountains and more locally in the Pyrenees Mountains to the north. Human-caused fires by contrast tend to be concentrated in the Erbo Basin, the agricultural region between the mountains. Here, much of the agricultural land has been abandoned and has been invaded by flammable herbaceous plants and shrubs. Reprinted from Amatullia *et al.* (2007) with permission from Elsevier.

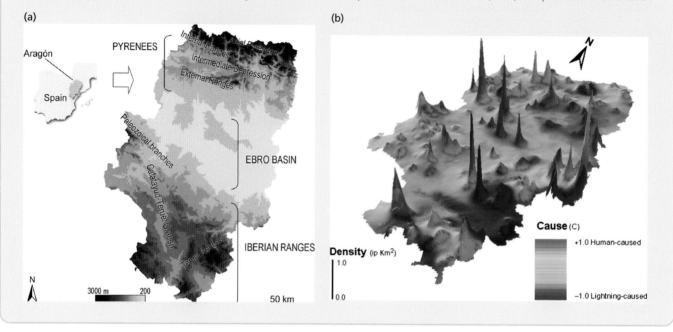

(a)

(b)

Where humans are clustered together it is inevitable that deliberate and accidental fires become more common. Thus, in Europe it is only the thinly populated countries like Russia and Finland, and tinder-dry countries like Cyprus where more than 10% of fires are started naturally. In others such as Germany, Turkey and around the Mediterranean basin, lightning-caused fires are down to just a few per cent. However, even in Europe, lightning-started fires can be important in sparsely populated areas – see Box 4.4 for an example. Similarly, in Alaska, Calef and colleagues (2008) confirmed that human presence has a positive effect on fire since they found that the number of fire ignitions increased near to settlements, roads and rivers where there are more people.

Cultural elements may play a major role in starting fires. In the north-west of Scotland, heather moorland (dominated by dwarf shrubs) is set on fire as a diverting pastime on a fine summer evening. Similarly in Spain, many fires are deliberate and attempts by the authorities to suppress them were met in the 1990s by the graffiti 'Wildland is ours, burn your share'. In densely populated Japan, where almost all fires are human-caused, 31% of fires are classified as being open fires or bonfires, many of which are fires started during the annual summer Festival of the Dead

(*Obon*). In this Buddhist ritual, family members hang lit paper lanterns or candles in cemeteries over several consecutive nights. Wind helps spread the flames into the surrounding landscape.

Human-caused fires also dominate in those parts of the world where fire is used as a common management tool. Africa stands out as a prime example. In Namibia, for example, fire is used to encourage grass for cattle grazing and to clear areas for cultivation, attract game, and clear around waterholes and honey-gathering areas. Fire is also used elsewhere in Africa for disease control – such as burning to control tsetse fly and mange tick populations. Of course, sometimes human fires take on an exaggerated role simply because natural fires are scarce. In Sri Lanka, for example, dry thunderstorms are rare and volcanic eruptions are absent, so the vast majority of fires are human-caused one way or another.

The overall conclusion is that globally, humans are now responsible for starting the majority of fires.

Which burns most area?

How much land an individual fire burns depends upon how continuous an area of fuel there is to burn and how quickly and successful suppression activities are. Perhaps not surprisingly, lightning fires in areas remote from people generally burn the largest areas. Between 1990 and 2000 in Australia, 26% of fires were started by lightning but they burnt half the area of all fires; by comparison, the 51% of fires caused by people burnt just 22% of the total burnt area. Similarly, in Canada, the 35% of fires started by lightning in 1994 resulted in 85% of the total burnt area, and in the USA, the 12% of lightning fires produced 54% of the burnt area. Conversely, where population density is higher and more uniform, and there is a good fire-management structure, lightning fires are more readily detected and do not contribute disproportionately large fire areas. For example, in Italy (see Box 4.5) the number of natural fires is much more proportional to the area burnt.

Box 4.6 illustrates that although some fire-causes result in proportionately larger areas burnt (as above), they can also have an effect on the size of an *individual* fire. In the Malaysian data given, a single case of arson made a significant contribution to the total area burnt in a 6-year period. This demonstrates that the statistics of area burnt can be frustratingly difficult to analyse – often a single very large fire or a single bad year will badly skew the data.

The fire behaviour triangle

Once a fire has started, its subsequent behaviour is determined by three things forming the sides of a triangle (Fig. 4.5): the amount and state of the fuel, the weather and the shape of the terrain (topography). We will deal with each of the three sides of the triangle in turn to see how they affect how a fire burns.

Box 4.5. Causes of fire in Italy in 1998. From Anon (2001)

Causes	No. of fires (%)	Total of area burnt (%)
Natural	1	< 1
Deliberate	51	74
Accidental	12	8
Unknown	36	18

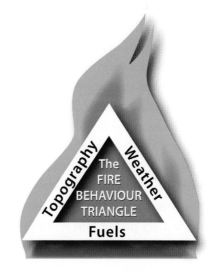

Figure 4.5
The fire behaviour triangle shows the three factors that affect the behaviour of a fire once it has started.

Box 4.6. Causes of forest fires in Malaysia (1992–1998). These figures do not include 65 000 ha of Sarawak that was cut and burnt by shifting cultivators. Many of the accidental fires resulted from throwing away burning cigarettes. From Anon (2001)

Causes	No. of fires	Total area burnt (ha)	Average area burnt per fire (ha)
Land-clearing fires spreading	16	3431	214
Unknown	8	556	70
Accidental	6	3384	564
Power transmission lines	2	25	13
Hunting	1	40	40
Arson	1	1000	1000

Fuel considerations: fires are what they consume

In the last chapter we saw how the chemical make-up, moisture and size and shape of pieces of fuel were important to the combustion process. Now we look at the bigger picture of how the burning of a fire is affected by the size, type, condition, arrangement, size and amount of fuel.

Types of fuel

In Chapter 3 we saw that small pieces of fuel (fine fuels, typically less than 5–6 mm in diameter) are most easily ignited and so play a major part in the initial stages of all fires and in the propagation of most of them. These are made up of dry grass, dead leaves, small dead twigs but also living leaves, needles and twigs especially if rich in flammable oils and resins. Once burning, the fine fuels act as kindling for the medium and the large pieces of fuel (more than 2.5 cm in diameter: branches, logs, standing dead trees – called snags – and stumps). These take longer to ignite, and so contribute to the fire only in large and intense fires. But they burn for longer and produce more energy than fine fuels.

We also saw that what the fuel is made of (how flammable it is) and how much water it contains also affects how it burns. So, for example, fire in a conifer forest will often burn around groups of less flammable and moist hardwoods. The drying of fuel is a key part of how well a fire will burn.

Drying of fuel

An essential difference between dead and live fuels is how they hold and lose moisture. Living vegetation will burn in a fire if there is enough energy to dry it out first (think of burning pruned vegetation on a roaring garden bonfire). But because of their high moisture, living material will usually act as a heat sink and retard fire spread and intensity. While dead fuel moisture is related primarily to weather conditions, moisture in live fuels is governed more by physiological processes. So, on a daily basis living plants tend to lose water during the day by evaporation and recover at night, but are buffered against too much loss and remain moister than dead vegetation unless there is a severe drought. How much moisture a plant will hold depends upon:

- What species it is (evergreen foliage has a lower moisture content than deciduous leaves but there is tremendous variation between even similar species and even individuals within a species).

- Whether it is annual or perennial (annual grasses tend to die off and dry rapidly during dry periods while perennials still have moisture-filled foliage).

- Season (new spring growth tends to be succulent and juicy, declining rapidly during the first few weeks and tapering off to the autumn).

- Age of the foliage (older foliage on evergreen species tends to increase in moisture during the summer and be lowest in spring and autumn).

- Amount of shade (shaded vegetation loses less water on a hot sunny day).

Dead fuels, on the other hand, lose and gain moisture in a much more passive way. They are hygroscopic, that is, they take up and give off moisture until they are in balance with the amount of water in the surrounding atmosphere. Moisture is absorbed as liquid or water vapour and lost by evaporation. In theory, fuels will approach an 'equilibrium moisture content' under constant conditions, but weather conditions are rarely ever constant and so the fuel moisture level is constantly following behind fluctuations in humidity. How rapidly fuels gain and lose water will depend on such things as what they are made of (leaves with or without waxy surfaces, or wood), size, location (in contact with the ground, exposure to direct sunlight and so on), weather conditions and even whether the fuel has already been dried before. Thus, although it sounds as if it should be fairly easily predicted, there are so many complex variables that on the small scale this is often very difficult to do. The safest generalisations are that:

- Thicker fuel will dry more slowly because (1) it can contain more moisture, (2) it has a smaller surface-area-to-volume ratio (see Box 3.4), and (3) evaporation from the surface is much more rapid than diffusion of moisture through solid fuel to reach the surface.

- Fuel moisture is usually lowest at mid afternoon, when it is warmest and lowest relative humidity, and highest at night.

Based on this, Byram in the early 1960s developed the concept of 'time lags' in the USA (Byram 1959). Dead fuels were classified by how long it took to lose two-thirds of their moisture under constant humidity and temperature. Once this time lag is known it will work for all conditions; i.e. if a fuel takes 4 hours to lose two-thirds of the moisture it can lose under set conditions, it will always take this time to come two-thirds of the way to equilibrium under any conditions. For very fine fuel this may be a matter of minutes, while for logs 15 cm (6 inches) in diameter it is around 36 days. Traditionally fuels (in the USA) were divided into classes of 1, 10, 100 and 1000 hour fuels corresponding to average sizes of 0–1/4 inch, 1/4–1 inch, 1–3 inch and more than 3 inches, respectively.

The importance of this sort of classification is that it allows you to visualise what is likely to burn in a fire. Thus after a dry day the finest fuels will burn, and after seven dry days (168 hours) material around 7.5 cm (3 inches) will burn and the fire is likely to be severe regardless of how much torrential rain may have fallen before. Similarly, the main effect of drought is not on the fine fuels but on the heavier logs, stumps and soil organic matter which will have time to dry and contribute to a fire. Thus, in the peat forests of Indonesia in 1997, extreme drought allowed normally water-logged peat to burn in enormous quantities, releasing between 0.81 and 2.57 Gt of carbon (a Gt is a thousand million tonnes) into the atmosphere, 13–40% of what humans produced across the whole world at that time.[1]

As an example of the importance of fuel moisture on fire behaviour, Van Wagner (1968) working in eastern Canada observed that in two similar fires in red pine (*Pinus resinosa*) but with differing needle moisture contents of 135% and 95%, rates of spread were 17 and 27 metres per minute with flame heights of 20 m and 30 m, respectively. In other words, drier foliage made for a more intense fire. A general rule of thumb is that crown fires are highly likely when the foliage moisture content is low – below 75% in Mediterranean shrublands and below 100% in conifer stands.

Fuel arrangement

The amount of fuel and how dry it is are important factors in fire behaviour but so is its arrangement in terms of its compaction (also called porosity – how close each particle is) and larger-scale spacing.

As anyone who has played with garden bonfires knows, fuel burns best when it is compact enough to maximise heat transfer between fuel pieces but open enough to allow enough air circulation to speed drying and bring oxygen in. For any fuel size, there is an optimum compaction; the larger the fuel pieces, the closer they need to be together. Logs burn best when piled together and touching and, unless very dry, will cease to burn if separated by much more than one diameter from each other. On the other hand, fine fuels burn best in loose arrays. The effect of compaction is nicely illustrated in these fine fuels. Litter from fir, spruce and many hardwoods that lies in solid compact masses burns very slowly even when dry whereas pine litter and

---

1. It was estimated in 1990 that anthropogenic emissions of carbon were 6.4 Gt per year. Between 2000 and 2005 this had risen to 7.2 Gt C per year (IPCC 2007).

curly deciduous leaves that form a light spongy mass full of air burn much more rapidly. Living leaves, spread out to collect light are often perfectly spread to burn, explaining why fires in grass spread more rapidly than in any other natural fuel.

Standing back a little, the spacing of fuel over a wider area, or its continuity both vertically and horizontally, is important for fire spread. Fire will spread across vertical gaps more easily than horizontal gaps because of rising heat but wind may alter this considerably (see Chapter 3 for a discussion on how crown fires start). So when fuels are even reasonably close vertically, fire will spread rapidly into the crown but as discussed in the last chapter, vegetation in forests tends to be layered, often with a large gap between the ground fuels (litter and shrubs) and the canopy.

Horizontal spacing is also important. When fuels are close together, the fire will spread faster but when fuels are patchy, scattered or separated by natural barriers such as rocky areas, streams or areas of bare ground, the fire will be irregular and spread more slowly. This is why the most common method of firefighting is to create a break in the fuel by physically moving it.

Just how big do fire breaks (breaks in fuel continuity) have to be to stop a fire? A small fire trickling through litter may be stopped by a gap of a few centimetres but large crown fires aided by fire whirls (see Chapter 3) have been known to cross a gap of 100 m to start a new surface fire the other side. If spot fires are involved then, as discussed in Chapter 3, fire can jump gaps tens of kilometres wide. Generally, as a fire gets more intense and bigger in area, it is less affected by minor fuel discontinuities and the fire spread mechanism becomes more efficient.

Amount of fuel

As the amount of flammable material increases, it is obvious that the amount of energy released by the fire also increases, producing fires that are more intense and difficult to control. But it is not just the overall weight of fuel that is important – the size distribution of what is there is also significant. Fine fuels provide the kindling and so determine whether a fire will start and spread but they have little influence on a fire's subsequent spread. Larger fuels provide the main source of energy and determine how intense the fire becomes (but bear in mind that large *living* fuels seldom burn at all and in fact act as heat sinks, slowing down the fire). Fire managers are thus very keen to know how much fuel there is.

Fuel load

The amount of fine fuel varies with time. Over a growing season, vegetation grows, dies and drops to the ground; litterfall varies from up to 3 tonnes per hectare (t/ha) each year in conifer forests to 6 t/ha in deciduous stands to 7–9 t/ha in tropical forests. Once on the ground plant material decomposes or builds up. Thus the amount lying on the ground can vary tremendously. In tropical areas, despite high litterfall rates, decomposition is so rapid that, with litterfall spread over the year, there may be little or no fine fuel on the ground. By contrast, in Australian eucalypt forests, fine fuel loadings (bits less than 6 mm in diameter) can reach 22–24 t/ha and in cool, moist conifer forests where decomposition is slow, litter can reach 300 t/ha. Bark shed by trees can also be an important contributor to fine fuels, especially in the eucalypts. In open southern eucalypt forests of southern Australia, for example, live shrubby fuels less than 4 mm in diameter contribute *c.* 4 t/ha while dead bark on tree trunks may add up to another 10 t/ha (Bradstock *et al.* 2002).

Normal figures for total fuel loading (everything that is physically capable of burning if given long enough) vary from 2–10 t/ha in grasslands to 70–150 t/ha in logging debris. Most forests will come somewhere between these two extremes. The size range of fuel tends to skew to the bigger end of things over time as logs build up on the forest floor. But even this varies. For example, Lee *et al.* (1997) found that log volumes in deciduous stands in Canada range from 109–124 m³/ha in Alberta in the west to 13–20 m³/ha in New Brunswick to the east; the east is damper with more rotting by fungi compared with dry Alberta, so logs last longer in the west and accumulate.

On top of all these gradual changes, the amount of fuel is abruptly changed by disturbances such as windstorms, fire, insects and disease, all of which can kill trees, plus human disturbance like logging. An increasingly worrying trend is the number of pines around the world that are being killed or seriously weakened by bark pine beetles of the genus *Dendroctonus*, including large areas of Canada, Siberia and even Central America (see for example Billings *et al.* 2004). In the southern 11 states of the USA it is estimated that the southern pine beetle (*Dendroctonus frontalis*) kills an average of 1.36 million tonnes of biomass each year (Coulson *et al.* 2005). Similarly, in British Columbia, Canada, the mountain pine beetle (*Dendroctonus ponderosae*) damaged 10 million ha or 68% of all the lodgepole pine forest (*Pinus contorta*) during the first 8 years of the century resulting in 40% mortality of the saleable pine timber, and a vastly altered forest structure (Axelson *et al.* 2009).

Available fuel

The above figures are *total fuel*, all the burnable material, dead and alive on a site. In some cases this can be guaranteed to completely burn; for example, in Australian bush fires much of the dead fuel is below 6 mm in diameter and is consumed by even low-intensity fires. In most cases, however, this is unusual and normally not all fuel will burn. For example, McCaw *et al.* (1997) looked at what burnt and what did not in a series of low- to moderate-intensity fires (less than 3000 kW/m) in debris left from thinning a eucalyptus forest (*Eucalyptus diversicolor*) in south-west Australia. They found that 95% of fine aerial fuels, 85% of 6–25 mm diameter fuel and about 55–58% of larger fuels were consumed. In other words, most of the dry fine fuels burnt but a significant part of the bigger (and moister) fuel did not.

Interestingly, only 56% of litter on the ground burnt but here the high moisture outweighed the fine nature of the litter.

How, then, do you decide what fuel is realistically likely to burn? Various theoretical categories can be put forward: 'potential fuel load', the amount of material that could be consumed in the most intense fire that could be expected to develop on a site; and 'available fuel load', the fuel which is available for combustion in a given fire (i.e. under specified fire weather conditions). All fine and good but in reality these figures can be very hard to predict, since the fuel consumed is dependent upon fuel moisture, size and condition. The amount of moisture in the fuel will be changing by the hour with larger fuels lagging progressively behind, as discussed above. Superimposed on this are seasonal changes in the amount and moisture of fuel. One day just the thinnest layer of litter might burn, yet later in the season a fire might involve the branches and duff, and even allow large logs to burn to just a line of ash. The fire itself may also alter the amount of available fuel. A fire may trickle through an area burning the available litter but drying out the living vegetation above. A change in wind direction and the area will be reburnt through the newly dried fuels. Or a fire burning over a deep but moist layer of litter (as in the Florida everglades) may remove the surface layer of dry litter exposing a blackened surface to the sun which may reburn a few days later, and repeatedly do so over the whole fire season.

There have been many attempts to quantify the amount of fuel using fuel classification schemes. The simplest is a direct estimation of the amount, usually described as low, medium, high or extreme, which has the merit of being easy once you have your eye in and at least helps decide which areas are likely to be a problem if there is a fire. It is also possible to mathematically model how much fuel there will be at a certain time in a certain forest type but it can be difficult to make the model accurately reflect real life. Another method used is to photograph a series of forest plots which are then dissected and the amount of fuel measured. In the field it is then a matter of matching the best photo to what you see.

Link between fuel and fire behaviour

The complex link between fuel and fire behaviour is nicely illustrated by a study carried out in the Canadian mixed-wood forest by Hély *et al.* (2000). They found that in computer model simulations, fires spread more slowly and were less intense in deciduous stands than mixed or coniferous stands. This was not primarily due to more fuel in the conifer stands (although they had significantly heavier loads of small-diameter twigs and shrubs) but rather the qualities of the fuel – combustible needles in a fluffy litter layer with lots of air spaces – and ladder fuels. Spring fires were more intense than summer fires, especially in stands with a higher hardwood component (before hardwood leaf emergence). Summer fires were less intense in deciduous stands because of the wet deciduous foliage and the cool, moist climate created on the ground. Spring fires were more intense in the same stands because the absence of deciduous foliage allows full sunlight to warm and dry the ground litter compared to below conifers.

The effect of climate and weather

In the section above and in Chapter 3 we saw that the amount and arrangement of fuel is important to how (and whether) a fire burns. However, while the fuel sets the stage, once out in the real world, it is the climate and the weather that can be of overriding importance in determining if fuel will burn and, if so, how much and how the fire will behave. Here we will define 'weather' as what happens day-to-day and month-to-month, and 'climate' as the pattern of weather over a longer period of years.

Climate effects

Climate has an overriding influence on whether fire is a naturally occurring event in an area. At its simplest, the burnable bits of the world have a climate with enough moisture to allow vegetation to grow (to create fuel), a dry season to create burning conditions and a source of ignition as discussed above. The less burnable bits will fail on one of these criteria. Thus, Malaysia (see Box 4.6) has fires but, on the whole, they have been small and infrequent – even including the comparatively large human-caused Sarawak fires, only 0.04% of Malaysia has been burnt per year on average between 1992 and 1998. Compare this to a similar-sized country such as Italy (Box 4.5) where 0.39% burnt each year on average between 1990 and 1999 – almost ten times as much as in Malaysia despite a large fire-suppression industry. Malaysia has an average rainfall of 2540 mm (compared with 657 mm in Italy), which together with a humidity normally exceeding 75% and a high rate of litter decomposition (which prevents fuel build-up) does not allow big fires. However, by changing these parameters, areas of Malaysia and other areas of South-East Asia's rainforest will burn; Langner & Siegert (2009) identify fire as the most important driver of forest loss in these areas. Fire itself in moist tropical forests is driven primarily by drought (Cochrane & Barber 2009) associated with El Niño years. In Borneo they found that 21% of the land area has been affected by fire. Most fires are aided by land clearance for agriculture allowing the sun in to dry the mass of dead vegetation (Fig. 4.6). Once fires start, they can gather enough intensity, helped by high winds, to be able to penetrate nearby moist tropical forest. Thus humans can override the natural influence of a poor climate for fire.

The effect of climate on fire is well illustrated by the effects of El Niño – changes in Pacific Ocean currents near the equator that occur at irregular intervals of 2–7 years and which tend to result in severe droughts leading to severe fires. The Ash Wednesday fires of eastern Australia in 1982/3, the headline-hitting fires of

Figure 4.6
A fire in the peat swamp forests of Central Kalimantan province on the island of Borneo has created fuel which will allow the next fires to burn further into the surrounding forest, gradually enlarging the area of disturbance. Forest fires during the 2002 dry season were widespread across parts of Indonesia, particularly in Central Kalimantan. Carbon emissions from fires in peatland forests are high owing to combustion of both forest biomass and surface peat. Photograph by Sue Page.

Indonesia in the late 1990s, and the big fires of southern Mexico and Central America of 1998 (where more than 10 000 fires consumed more than 4000 km^2 in 3 months in southern Mexico alone) all coincided with El Niño droughts. The effects of El Niño in the Pacific are even felt in North America, particularly from Arizona and across to Florida. In this case, however, El Niño years paradoxically (at first sight) result in smaller areas being burnt. The reason is that El Niño events tend to affect the weather most strongly in the North American winter months and weaken or disappear before the summer fire season, and, if anything, produce a wetter spring.

Normal large-scale variations in climate from year to year also have a lot to do with how much forest is burnt each year over large areas. For example, the development of big weather systems over western Canada (particularly persistent high pressure ridges that block cloud movement) have been shown to coincide with large fire years in the Canadian Rocky Mountains. However, the effect of regional climate on fire regime is not always obvious. For example, in northern Sweden fires became more frequent around 2500 years ago despite a change to an overall wetter climate. This was undoubtedly because there were more frequent dry summers hidden within this wetter period, leading to more fire.

Weather effects

If fuel load and climate set the stage for fire, then it is weather that writes the script. Weather controls the drying of fuels and plays a crucial role in subsequent fire behaviour. Weather conditions can be the deciding factor on whether a fire is a gentle trickle through the litter or a raging inferno driven by low moisture in the fuel, warm dry air and high winds. This is of critical importance to the firefighting industry; a fire is most likely to be hard to put out when it has the potential to spread fast and intensely. It is often the case that much of the area burnt in a year occurs on just a few days of severe weather – a common axiom is that 97% of the area burnt is consumed on just 5% of the days. This is why so much time and money has been invested in many parts of the world trying to link weather and fire, so that the risk of fire can be more accurately predicted. The most important features of weather are precipitation, wind and temperature, and the effect of these three on humidity.

Precipitation and dry spells For fire to burn there is a need for dry fuels, usually with less than 15% moisture (i.e. less than 15 g of water for every 100 g of dry fuel). Fuel moisture is obviously tied to the amount of rain and its rate of evaporation. There is often a good correlation between the amount of spring and summer rain and area of a landscape burnt in a year. But really it is not so much the total amount of rain that falls but the timing of it, particularly the gaps between rain showers that give the fuel time to dry. In two hypothetical identical areas, both receiving 300 mm of rain over the summer, there should be more fire in one that gets torrential rain just once a month than the one that gets a misting every day or so. In Canada, Flannigan & Harrington (1988) looked at three decades of records and found exactly this – severe fire months were not dependent upon amount of rainfall but were strongly influenced by the length of dry periods – the longer the dry period, the bigger the fires. The most important predictors of severe fires were found to be long sequences of days with less than 1.5 mm of rain and long sequences of days with relative humidity below 60%.

The effectiveness of rain at wetting a fuel is linked to how long the rainfall lasts. A shower delivering 10 mm of rain can much more thoroughly wet the fuels if it falls over several hours rather than just several minutes; short heavy rainfalls leave little time for soaking in, and produce a heavy run-off. Large fuels in particular, like logs, will tend to shed rain so absorption of water is dependent upon duration of rain.

As discussed above, only a short period of hot and dry weather can dry out the fine fuels to a state where they can carry a fire. In Canada it is generally accepted that only three days of hot (25+ °C) and dry (a relative humidity less than 30%) weather is sufficient to dry out the fuel enough to sustain combustion.

Rain is important not just for wetting fuels but also for its production in the first place. Certainly in Mediterranean Europe (and undoubtedly elsewhere), growth of fine fuels each year depends very much on the amount of rainfall during the previous winter–spring season producing fuels for the summer. It has been found in Portugal, for example, that as rainfall in January–April (the main growing season) increases, the vegetation grows more luxuriantly, producing more fuel, resulting in more being available to be burnt. But eventually the relationship changes as excess rain creates a large moisture reserve in the vegetation and soil, which hampers burning in the summer, reducing the area burnt.

Wind This is also of paramount importance in how a fire spreads. The moderating effect of moisture in fuel can be effectively neutralised when a fire is driven by strong winds – fire breaks or six-lane roads can become meaningless. Fires can be divided into two basic types: convection-column driven and wind driven.

The convection column, described in Chapter 3, is formed where the upwards convection currents of the fire are stronger than the wind and take over as the main source of air movement. As the convection column builds it can dampen or resist minor fluctuations in wind direction. The column pulls in air from surrounding areas, producing a normally slow-moving fire but one that has great destructive power. High winds can shear a forming convection column as the wind speed increases with height above ground (due to less friction with the ground).

In a wind-driven fire, the horizontal wind is powerful enough to prevent or break up the convection column. Fire spread is greatly enhanced as the wind directly affects spread by pushing over the flames to heat the fuel ahead more effectively, blowing the heat of the fire through the trees ahead to preheat and dry burnable leaves and twigs, mixing in more oxygen and carrying burning embers ahead to create spot fires (see Fig. 4.7 and also Figs 3.4 and 3.9). Fires can accelerate in sudden spurts as the wind shifts or gusts. Under milder burning conditions, fluctuations in

Figure 4.7
Wind helps a fire to spread more rapidly. The bent-over flames help to dry and preheat the burnable leaves and twigs in the trees ahead, making them easier to ignite. The wind also mixes in more oxygen and can carry burning embers ahead to start new fires a long way upwind. From Anon (1970).

wind speed and direction are one of the major factors in determining the rate of acceleration. Fuels composed of fine particles like dead needles (rather than twigs and branches) are particularly sensitive to wind, and the rate of spread can vary a hundred fold with strong winds compared to no wind.

Examples of the devastating effect of wind are numerous. In 10 days in autumn 1993, more than 80 000 ha of dense chaparral in southern California were burnt in an area 150 × 80 km. The fires were helped immeasurably by the intense Santa Ana winds bringing dry hot air from the Mojave Desert towards Los Angeles (Keeley 1998). The fires easily overwhelmed suppression attempts and in a short time destroyed around one thousand homes at a cost of US$ 1 billion (Minnich 1998). Nor is this atypical; many areas of the world have their own hot, dry winds particularly associated with extreme fire danger: the dry and turbulent Mistral in southern France, the Tramontana in northern Italy, the Föhn wind of the Alps of Germany and Switzerland and the Chinooks of the eastern Canadian Rockies.

Other dry winds can be formed by patches of super-dry air forming a 'dry slot' some 2–6 km above the ground. On very hot days in Australia and the USA (and undoubtedly elsewhere) convective currents can rise above 5 km and result in this super-dry air being brought to ground level. When this happens, relative humidity can drop from 20% to around 3% for 2–3 hours at a time, more than enough time for the fine fuels to dry. These dry conditions are also linked to gusty wind conditions, potentially resulting in a very explosive situation should a fire start. In southern Australia, the extreme fire on the Eyre Peninsula in January 2005 has been linked to a dry slot. This fire burnt more than 77 000 ha causing some AU$27 million of damage and caused the loss of 93 homes, 46 000 animals and the death of nine people. Mills (2005) gives other examples.

(a)

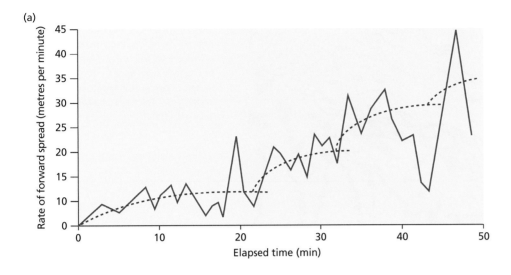

(b)

Less dramatic but perhaps more dangerous is the change in wind direction which can cause a relatively benign fire flank to roar into life. Wind shifts have the effect of rapidly increasing fire area and potentially catching firefighters unaware. In Australia (Fig. 4.8a), sudden increases in rates of fire spread have been linked to changes in wind speed and direction. Generally a fire develops into an ellipse spreading along the path of the wind. As the wind changes direction, what was the slow-moving flank of the fire becomes the new front, now much broader than before (Fig. 4.8b). The same pattern has been observed in the eastern USA. Brotak & Reifsnyder (1977) found over three-quarters (78%) of large fires (defined as fires larger than 2000 ha) between 1962 and 1973 were associated with the passage of a cold front, and usually a dry one at that. A cold front means that a new dominant airmass has moved into the area, and cold fronts can move quickly and produce rapid changes in wind speed and direction. Unfortunately the term 'cold front' does not always herald cold conditions – the temperature change could only be a few degrees. In the northern hemisphere surface winds shift from SW to NW with

Figure 4.8
Sudden increases in speed of spread of a grassland fire in Australia are shown in diagram (a) as the dotted lines (redrawn from Cheney 1981). These are caused by sudden shifts in wind direction. The pictures in (b) show an extreme case of this happening. A fire blown by the wind has a narrow fire front with an ever-increasing flank length. With a sudden change in wind direction, the flank becomes a very broad fire front and the fire roars across the landscape.

the passage of a cold front. This 90° direction shift turns the wide fire flank into the new front, that, with the high winds generated by the front, results in more vigorous burning and a fire more difficult to control and likely to grow large quickly.

Temperature If we want to dry things, such as clothes, we put them in a warm, dry place because the heat aids evaporation of water. In the same way, higher air temperatures help fuels to dry – especially the thinner fuels that respond most quickly to weather changes (see Chapter 3). It could be argued that warmer fuels also need less energy to raise them to the temperature of ignition but in reality a difference in fuel temperature of 10 °C either way is unlikely to make a significant difference considering that the fuel must be heated to more than 300 °C for ignition.

While ground-surface temperature has a direct impact on the fuels, just as crucial to fire spread are the temperature changes in the column of air high above a forest. If you go up in a hot air balloon, the air gets noticeably cooler, partly because you are further away from the source of heat – infrared radiation from the ground – but also because the air gets thinner (less dense) and as gas expands it cools (this is termed the adiabatic lapse rate). Air temperature normally drops by 0.7 °C for every 100 m height gain (3.8 °F per 1000 feet) but this can vary. If the air temperature gradient is steep, with the ground being much warmer than higher up, this creates 'unstable' air where vertical air movement is encouraged. In 'stable' air, this effect is much weaker. During the summer months, you can tell the air is unstable if you observe afternoon cumulonimbus clouds – or thunderstorm clouds. These are formed when air is heated at the surface of the earth and it rises rapidly and the moisture condenses into clouds; typically the bigger (taller) the clouds, the more unstable the atmosphere. Unstable air tends to increases the height and strength of convection columns above a fire (see Chapter 3) leading to fiercer burning, and can lead to winds from higher up in the atmosphere swooping down to ground level producing stronger and gusty surface winds. As noted above if these winds are associated with 'dry slots' this can lead to very severe burning conditions.

Temperature inversions (where it is *warmer* as one goes higher) act as a barrier limiting vertical movement of air. Smoke will rise to the inversion and pool underneath the invisible ceiling, spreading horizontally. Night-time inversions, caused by the ground cooling, usually disappear by midday so fire activity can increase as the air becomes increasingly unstable. This can lead to dangerous situations as when, for example, inversions in valley bottoms have ameliorated fire behaviour during the morning hours, and unsuspecting firefighters are caught off guard when the inversions break up and the fire acelerates rapidly.

Winds can also cause some erratic fire behaviour such as fire whirls, horizontal roll vortices, spotting (see Chapter 3) and development of pyrocumulus (thunderstorms that produce abnormally large amounts of lightning).

Humidity Humidity is water vapour present in the air, expressed as a percentage of how much water the air can hold at a given temperature. Warm air absorbs more moisture, and so for a given amount of water, will have a lower humidity and can absorb proportionately more. So humidity gives a good indication of the evaporative power of the air. Air is usually 'drier' during the day, when it is warmer, than at night. Fires normally burn more slowly at night because the fuel will absorb moisture from the damp night air.

Fire season

Cyclical variations in yearly weather and fuel production combine to create the fire season: the time of year when most fires occur naturally. It is usual to assume that this will be at the hottest and driest time of the year – usually the summer – but there are variations. In the north of Australia fires are most common from April to November (the dry season) while in the south the main fire season is in summer from November to March. In the west of South Africa fires in the fynbos shrublands are most prevalent during the summer dry months while late summer and autumn fires are frequent in the southern Cape. In the north of North America, the fire season tends to be primarily in the spring (May and June) with a second blip in the late summer: in spring once the ground has dried but before there is too much new growth of moisture-filled green vegetation, and in late summer once things die off and dry but before autumn/winter wet and cold comes. Further south the fire season also covers the summer and further towards the drier and hotter south it starts progressively earlier and finishes later, becoming almost year round. In general in temperate and tropical regions, the fire season lasts 3–4 months while in more arid areas, the season last 6 months or more.

Variability in how a fire spreads

A fire rarely burns evenly in all directions, producing a circle around the source of ignition. We have already seen that wind can make a big difference, driving a fire into an oval. But in practice it is usually even more uneven with an irregular shape, with some areas burnt more intensely than others, or not at all. The result is a mosaic of different patches. Why does this happen?

Chance

The first reason is really just chance. Vagaries of wind blowing the fire in different directions can result in what seem to be identical areas being burnt in different ways, or escaping fire altogether.

Fuel quality and amount

Fuel in a forest is rarely ever completely homogeneous. As described in the previous chapter, fuel varies in quality (chemical composition and moisture content) and in

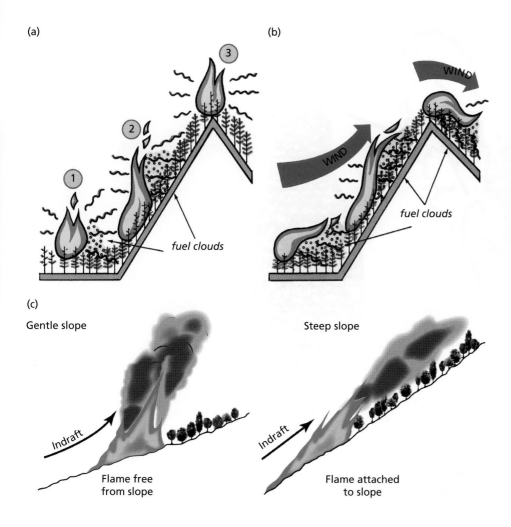

(a)

(b)

(c)

Gentle slope

Steep slope

Figure 4.9
(a) Fires 1 and 3 tend to preheat relatively little fuel since the burnable leaves and twigs are not that close to the energy radiating out from the flames, but with fire 2 the flames are closer to the fuel on the uphill side, and the fuel is heated, dries and ignites more easily than if on level ground helping the fire spread uphill faster than on the flat. Add in wind up the slope (b), bending the flames towards the fuel and the preheating is even greater (even on the flat) and the fire burns even quicker. (c) Convected heat rises along the slope, often clinging to the slope, causing a draft which further increases the rate of spread. (From Cottrell 2004.)

amount (how much has accumulated, how much is dry enough to burn) and arrangement (how the bits of fuel are arranged and thus how much heat is transferred from one piece to another, and how much is lost). These variations are especially important under gentle weather conditions, where strong winds or high temperatures are not overpowering influences.

At the landscape scale, such variations in fuel may reflect pockets of different species growing. For example, areas of hardwoods will burn differently from areas of conifers.

Topography

Topography is the shape of the landscape and is usually broken down into three main components – slope, aspect and terrain. All these can have important (and complex!) effects on the rate and direction of fire spread (Fig. 4.9).

Slope is the steepness of the land and has the greatest influence on fire behaviour since it affects both the direction and speed of the fire. Fire tends to burn faster uphill than downhill, and so a fire usually moves up a slope. The steeper the slope, the faster

Figure 4.10
Steep valleys or gullies are critical since they can create a chimney effect, causing a forced draft, as in the chimney of a house. The rapid drawing-in of fire-warmed air from lower down the valley and the preheating of fuel results in an explosive situation where fires spread quickly and intensely; situations notorious for claiming lives. From Wilson & Sorenson (1978).

the fire will move, and as a rule of thumb, the speed of a fire doubles for each 10° increase in slope. Steepness works this way for several reasons, as explained in Figs. 4.9 and 4.10.

It is perhaps not at all surprising that most housing losses from fire in shrub and forest are associated with steep slopes facing the prevailing winds, encouraging the fire to move upslope often with great speed.

The speed of a fire moving down a slope is almost independent of how steep the slope is. This is because the fire moves mainly by the radiation from inside the glowing fuel heating the next bit of fuel, rather than the heat from the flames. In fact, a fire burning downhill with no wind will have a somewhat similar behaviour to a fire burning against the wind simply because the angle between the unburnt fuel and the flames will be similar.

Aspect is the compass direction that the land faces. In the northern hemisphere, south-facing slopes are pointing more directly at the sun and so receive greater amounts of direct heat from the Sun. The vegetation also tends to be thinner (because it is drier) letting more heat reach all parts of the fuel and allowing air to circulate freely, so living and dead material dries quicker. Heating by the sun also causes earlier and stronger slope winds. So the combination of higher temperature, stronger winds, lower humidity and lower fuel moisture results in south and south-western facing slopes (in the northern hemisphere) being a greater fire hazard. These are the conditions needed for easy ignition and a rapid rate of fire spread.

Terrain, or shape of the land, plays a large role in fire dynamics by providing barriers to fire spread. Regions with more fire barriers (such as scree slopes, cliffs, wetlands and rivers) tend to have smaller fires and so parts of the landscape burn less frequently (e.g. Iniguez *et al.* 2008). Terrain also affects fire by controlling wind patterns. Wind flows like water in a stream and will be deflected by barriers, speeding up and slowing down as it goes around and over obstructions. For example, wind going over a hill has a longer path to follow and so speeds up. This is felt least in the protected angle at the base of the hill and most on the higher slopes and top. At the ridge top, the wind eddies over the top (a 'roll eddy') which may be exacerbated by an 'upslope airflow' from the other side of the ridge, causing turbulence. This slows fire spread but the turbulence can pick up burning embers and shower them on the opposite slope causing spot fires. Other obstructions such as trees and rocks, or sharp bends in valleys can also cause turbulence or eddies, affecting the direction and speed of a fire. Terrain is also important in terms of fires starting from lightning strikes, which are expected to hit high points, summits and convex surfaces more than elsewhere.

As wind flows through a restriction, such as a narrow ravine, it increases in strength, with a direct effect on the fire, as in Fig. 4.10. The steep terrain can add to the problem since heat radiation can dry and preheat the slopes (especially the one with most sun on it) making it liable to explode with fire if a spark hits it. In the shrubby chaparral of south-western USA it is said that this preheating of a slope and evaporation of flammable gases is like adding around 10 000 litres of petrol per hectare (1000 gallons per acre). In this way a whole hillside can ignite almost instantly.

On the other hand, barriers in the terrain such as rocks, lakes, bare moist soil, wetlands, roads, changes in fuel type and previously burnt sites can stop a fire from spreading. Indeed firefighting will often try to produce such a barrier (by physically removing fuel or making it too wet to burn).

The complexity of fire spread

Dissecting how each component of the land's shape affects fire behaviour, as we have done above, is fairly intuitive but when it is all put together, it becomes much more difficult to predict exactly how a fire will move over a particular piece of land under particular weather and fuel conditions. And it all changes with time as the day warms and winds vary, and as weather fronts move through. To predict what will happen to a fire is the stuff of years of experience. So when the inevitable happens and things go wrong and firefighters lose their lives, we should spare a thought for the huge and complex undertaking faced by fire managers.

Patterns/mosaics on the landscape

From the above it can be seen that fire is not a uniform agent of change on the landscape. Areas will burn at different intensities; some areas may even reburn more than once during a fire as fuels dry. The result is a mosaic on the landscape. These range from small-scale variations (different amounts of burning over just a few centimetres to a few metres) to the large scale (such as the typical pattern of burning on a hillside) or to the landscape scale where, for example, a particular valley may be less burnt due to high humidity from a river.

Patterns of fire over time

So far we have been dealing with one fire and how it burns. Now we need to consider a wider perspective and look at multiple fires over time. This is the last piece in the 'fire regime' of an area, defined as the characteristic frequency, extent, intensity, severity and seasonality of fires within an area.

Fire frequency or fire interval are normally considered to be the most ecologically useful way of measuring repeated fires over time (see Box 4.7) because they relate directly to the life history of the plants and animals that inhabit a forest. Having said this, it should be noted that it is not just the average interval between fires that is important but also the variance of fire intervals, since longer or shorter breaks between fires may have significant effects on what does and does not survive a fire.

Typical fire frequencies in different forests are given in Box 4.8. This says nothing about the type of fires that would be expected each time which will obviously depend upon many factors. Generally, however, there is a tendency for large, high-intensity crown fires to occur in forests with a low fire frequency (there is plenty of time for large quantities of fuel to accumulate). Conversely, relatively small, low-intensity surface fires tend to occur in forests with a high fire frequency since only a small amount of fuel accumulates between each fire. Another way of saying this is that fire intensity and size tend to be inversely related to fire frequency.

Fire frequency can also change with time. In the sequoia groves of California, Swetnam (1993) found that during periods of high fire frequency, fuels did not have much time to accumulate between fires resulting in a patchy pattern of smaller fires. During periods of low fire frequency, however, more fuel accumulated and the resulting fires were bigger and more intense, producing a larger, more uniform distribution of vegetation and fuels (a 'coarse-grained' spatial pattern). In this case, the fluctuation in fire frequency was caused by variations of climate over hundreds of years (see below), but the same patterning has been achieved

Box 4.7. Different definitions of how repeated fires are measured over time. These terms have sometimes been wrongly used interchangeably – be careful!

A hypothetical example. Five fires (numbered 1 to 5) have burnt in a forest over a 200-year period. The dates between the arrows are the times in years between each fire.

Fire frequency (fire occurrence): the average number of fires that occur per unit time at a given point. Thus the fire frequency in our example is five fires in 200 years (although for convenience this is normally spoken of as an average fire frequency of 40 years).

Fire interval (fire return interval, fire-free interval): time in years between two fires in a given area. In the example forest, the interval between fires 1 and 2 is 56 years, and between fires 3 and 4 it is 20 years. It is important that the size of land area this applies to is specified.

Fire period (mean fire interval): the average of fire intervals in a given area and a given time span. In our example, the fire period is 50 years in this forest over the 200-year period. The time span and surface area it applies to should be specified. In many ways this is very similar to fire frequency but most tightly prescribed.

Fire cycle (fire rotation period): the number of years required to burn over an area equal to the area of interest. Size of the area of interest must be stated. Thus, if our example forest was 200 ha, and the first fire burnt 100 ha, the second fire 25 ha and the third fire burnt, for convenience, 75 ha, then the fire cycle would be 105 years. This definition does not imply that the entire area will burn during a cycle; some bits may burn several times and others not at all.

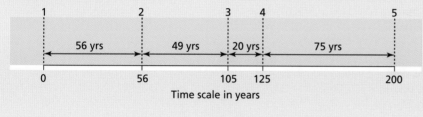

elsewhere by human fire suppression. Minnich (1983) found a fine-grained pattern in the chaparral shrublands of northern Mexico where frequent small fires are allowed to burn, but across the border into California where fire suppression was more active and so most significant fires were large (the small ones were readily put out), the landscape had a more homogeneous (coarse-grained) structure. Similar patterns are seen in ponderosa pine (*Pinus ponderosa*) in south-west USA where 70 years of fire suppression has shifted pre-settlement fire regimes from frequent low-intensity surface fires to infrequent, but increasingly numerous, large catastrophic crown fires.

What affects how often a piece of land reburns?

First of all, chance plays a part in whether lightning or other ignition sources happen to coincide with fuel condition and weather conducive to a fire, and therefore which year an area will burn.

Box 4.8. Fire frequencies (see Box 4.7 for a definition of this term) in different types of forest and bush. Based on Whelan (1995, p. 53) and Chandler *et al*. (1983)

Forest/bush type	Place	Fire frequency (years)
Boreal forest		
Spruce	Sweden	Up to 500
Moist spruce/hemlock/pine, E. Canada		340
White spruce (*Picea glauca*)	Flood plains of Alaska, Yukon	200+
Jack pine (*Pinus banksiana*)	Northwest Territories, Canada	25–100
Lodgepole pine (*Pinus contorta*)	British Columbia	50
Scots pine (*Pinus sylvestris*)	Sweden/Finland	45–100
Sub-alpine forest		
Lodgepole pine	Montana	25–150
Lodgepole pine	Sierra Nevada, California	100–300
Balsam fir (*Abies balsamea*)	New England	1000+
Moist temperate forest		
Southern beech (*Nothofagus* spp.)	Tasmania	300
Eucalyptus	Tasmania	100
Dry temperate forest		
Longleaf pine (*Pinus palustris*)	SE USA	3–20
Mixed conifer forest	California	1–15
Giant sequoia (*Sequoiadendron giganteum*)	California	10–100
Aleppo pine (*Pinus halapensis*)	E Spain	50
Mediterranean-climate bush		
Evergreen chaparral	California	20–50
Deciduous chaparral	California	30–100
Semi-arid deserts		
Desert scrub	Arizona	50–100
Pinyon pine – Juniper (*Pinus monophylla* – *Juniperus* spp.)	W USA	100–300
Tropical forests		
Moist evergreen scrub	Florida	20–30
Rainforest	Equatorial areas	never?

Second, rate of fuel accumulation will also usually determine when a fire is next likely to successfully spread. After a fire there may be very little left to burn, and living and dead material will accumulate over the years until it normally reaches a maximum. The quicker fuel builds up, the sooner a fire can be sustained. Thus the fynbos shrublands of South Africa need 4–6 years to accumulate sufficient fuel to burn, and fires occur at random within a 6–40 year rotation (Booysen & Tainton 1984). The amount of fuel will also help determine the size of the fire since the more fuel there is, the more likely that it will be continuous and that there will be sufficient heat to help jump any remaining gaps. This affects fire frequency directly since the larger each fire is, the more likely it is that the whole area will be reburnt by successive fires.

Finally, as already discussed, suitable weather for growing fuel has a great effect on whether (and how) a fire burns. Thus, the occurrence of suitable years can have a profound effect on fire frequency. For example, the Karoo of South Africa rarely accumulates enough fuel to support a fire, but following unusually high rainfall seasons, they can and do burn.

In the longer term, changes in climate can have an equally important effect on fire frequency. For example, Clark (1988) looked at fire scars and charcoal deposits (see below) and found that fluctuations in climate over the past 750 years corresponded to changes in disturbance regimes – fire frequencies reduced dramatically with the onset of the little Ice Age around 1600, compared with the warm dry period preceding it. Swetnam (1993) found a similar story. He looked at fire scars in giant sequoias (*Sequoiadendron giganteum*) and reconstructed the fire history of five groves of trees during the past 2000 years. He observed that fires were frequent and small during a warm period about AD 1000–1300, and less frequent but larger during cooler periods from about AD 500 to1000 and after AD 1300. It is generally the case in these reconstructions of fire history that fire frequency (e.g. number of fires per century) is similar over large regions at the same time, adding credence to this being an effect of large-scale climatic processes.

The current global climate change is having a similar effect. Fires are becoming more frequent and, due to major shifts in drought, areas such as tropical rainforests are seeing fire for the first time in millions of years.

Reconstructing fire history

It is worth expanding on just how fire histories are worked out. This information is often needed to help decide what future management should be, especially as fire frequency can have dramatic effects on what trees make up a forest (see Chapter 5).

Historical records (oral and written) are valuable. For example, historical documents from forest companies and protection agencies have been used in producing a fire history for eastern Canada (Bergeron *et al.* 2001). In Europe, records from land survey maps, court records, annual reports from district forest officers, records of land inquiries and travel descriptions have been used to reconstruct fire histories in the forests of north Sweden. Old photographs of stands can show past fires, but care is needed since recent fires tend to wipe out evidence of older fires.

Many people have used the oldest trees on a site as an indicator of when the last fire burnt through, although this may miss very low-intensity fires. However, trees can contain direct evidence of past fires in the form of fire scars (see Box 4.9 and

Box 4.9. Prehistoric fire scars from White Moss in the English Midlands

White Moss was a small peat bog in central England which in the mid 1990s was excavated to get at the commercial amounts of high-quality lake sand beneath the peat. As the machinery dug down, a layer of Scots pine (*Pinus sylvestris*) stumps was found between the sand and the peat as in (a). These remains represent a forest that grew on the newly infilled lake some 5000 years ago. The peat had preserved the stumps although not before insects and fungi had caused some damage, such as can be seen in the cross-section in (b).

It was noticed that a number of the stumps had fire scars – as in (b) and (c). Radiocarbon dating of the wood was possible but is not completely accurate. Fortunately, the pattern of wide and narrow rings through the stumps could be matched against a tree ring chronology that goes back from the current day, so it was possible to give calendar years to each fire. Moreover, because the scar is midway through a ring, it is known that both fires burnt in late spring. In this case, the fire in (b) was in late spring 2800 BC and another on the site (c) was in late spring 2710 BC (Chambers *et al*. 1997). Photographs (b) and (c) by Gerald Burgess.

(a)

(b)

(c)

Figure 4.11
A fire-scarred ponderosa pine (*Pinus ponderosa*) from Arizona. The centre of the tree is dated at 1648 and met 16 fires, the last being in 1943. Note that the fires recorded in the fire scars are all on the same side of the tree except the fire in 1886 that came from a different direction. See also the burnt tree in Fig. 5.1.

Chapter 5 for details of how they are formed), which tend to be cumulative. Since temperate trees produce one growth ring of wood per year, old scars can be accurately dated, and trees may contain evidence of dozens of fires (Fig. 4.11). Dieterich & Swetnam (1984) recorded 42 scars in one 327-year-old ponderosa pine!

Ninety giant sequoias sampled by Swetnam (1993) had an average of 64 fire dates recorded in each tree, and he was able to trace a fire history back 2000 years. But do trees record all fires that pass the tree? Not necessarily – just those that are hot enough to damage the living tissue of the tree, more specifically the cambium. It seems that the resin that tends to ooze out onto the 'catface' of a fire scar increases the likelihood that the next fire will ignite the tree, but if the scar has a chance to completely callous over, it may develop an immunity to subsequent scarring because of the increasingly thick layer of fire-resistant bark. And really hot fires may burn sufficiently far into the wood to remove old scars. Having said this, Swetnam (1993) is convinced that giant sequoias at least hold a very good fire record:

> Even low-intensity surface fires burning up to the base of sequoia trees often
> radiate enough heat to kill living tissue along the edges of old fire-scar

cavities where the bark is relatively thin. Because of the low flammability of sequoia wood and resin, repeated fires usually do not burn off the lesion (scars) caused by older fires.

Indeed, fire scars have been the most important source of fire history information around the world – Arno & Sneck (1977) give a very readable account of how to go about working out fire history from fire scars.

Sediments that build up in lake bottoms and the peat accumulating in bogs have also yielded valuable information of fire frequencies. Charcoal particles carried in the air or washed by water are caught in layered sediments and peats which can often be assigned dates (see Cope & Chaloner 1980, Patterson *et al.* 1987). Associated pollen can also yield information on vegetation changes that go with an opening of the canopy and subsequent regrowth. Lake and reservoir sediments have also been studied for unusual or extreme run-off events that are indicative of vegetation removal.

A major limitation of fire scars and sediment/peat records is that they often give little spatial information. A fire-scarred tree says little about how big the fire was unless there are a large number of scarred trees that can be used to outline the limit of each fire. Similarly, charcoal in sediments and peats may have been transported long distances in the air before being incorporated and so give even less information on just where and how large the fire was (although larger charcoal fragments are likely to be more locally produced – a good review is given by Butler 2008). Similarly, intensity of the fire and the season it burnt in are usually lacking (although fire scars can sometimes be linked to a season – see Box 4.9).

Fire size – how big is big?

In southern Europe, Viegas (1998) suggests that 0.01 to 10 000 ha (100 km^2) is the typical range, and this seems a fair rule of thumb elsewhere. He proposes that 1000 ha (10 km^2) is the general limit between a medium and large fire, while Gill & Allan (2008) suggest that 'huge' fires are those over a million hectares (10 000 km^2). Large fires have occurred across many parts of Australia and on other continents. A fire in Quebec, Canada was almost 500 km^2 (Fig. 4.12) and fires in western USA and northern Canada have been in excess of 2000 km^2 and may represent a maximum that is likely under current fuel and climatic conditions in these areas. The bush fires of Australia collectively have reached over 550 km^2 in some years. In North America the biggest recorded fire – the Miramichi fire of October 1825 – burnt 120 000 km^2 (1.2 million ha or 3 million acres) of New Brunswick, Canada and Maine, USA (see Chapter 2). Large fires are started by single or multiple ignitions and usually become large because of a combination of fast rates of spread, sustained favourable burning conditions and failure of initial fire suppression. Large fires are a problem since if suppression efforts are prepared for the average

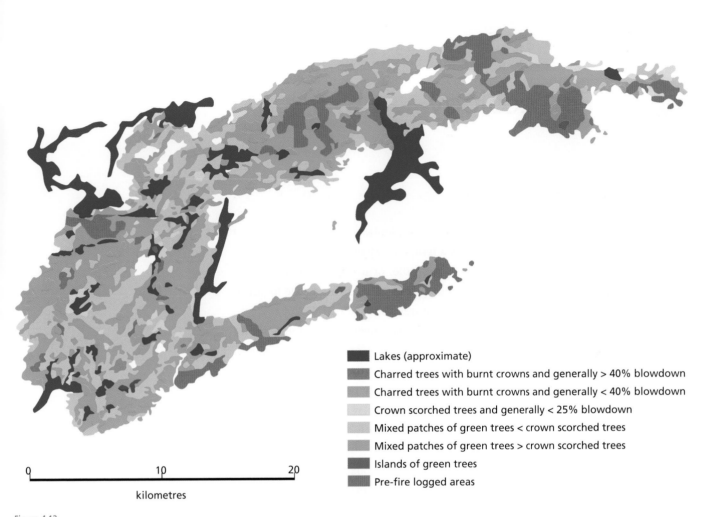

Lakes (approximate)
Charred trees with burnt crowns and generally > 40% blowdown
Charred trees with burnt crowns and generally < 40% blowdown
Crown scorched trees and generally < 25% blowdown
Mixed patches of green trees < crown scorched trees
Mixed patches of green trees > crown scorched trees
Islands of green trees
Pre-fire logged areas

Figure 4.12
This map represents a large fire in western Quebec, Canada in August 1995 that burnt almost 500 km^2 (49 070 ha) of boreal forest dominated by black spruce (*Picea mariana*) and jack pine (*Pinus banksiana*). Parts of the area had previously been burnt in 1820, 1870 and 1905 and some areas had been recently logged so there was a pattern of trees of different ages on the landscape before the 1995 fire. As can be seen, the fire left behind a complex mosaic of different areas. Charred trees with completely burnt crowns occurred in 108 patches (average size 194 ha) covering 43% of the total area burnt while 3% of the area was completely missed by the fire. © International Association of Wildland Fire 2001. Reproduced with permission from the *International Journal of Wildland Fire*, **10**(2): 119–127 (Kafka, V., Gauthier, S. & Gergeron, Y.), http://www.publish.csiro.au/nid/114/paper/WF01012.htm. Published by CSIRO Publishing, Melbourne Australia.

situation, then they can deal with double but not ten or a hundred times that size. Fires are too big for effective control when rate of perimeter growth exceeds rate of effective suppression.

Average fire size is a rather meaningless concept since larger fires account for most of the area burnt: as another rule of thumb, the top 3% of fires account for 97% of the area burnt (Stocks *et al.* 2003).

5 | Fire ecology

When considering forest fires and the survival of plants and animals there is a glaring paradoxical imbalance. A flame has a temperature of around 800–1200 °C whether it is the gentle flame of a match or candle, or a raging forest fire (see Chapter 3 for a longer discussion), and physiologically active living tissue (plant or animal) is killed by a short exposure to temperatures above 50–60 °C.

A burning forest does not consist of solid flame, so inevitably the temperature inside the forest is not uniformly that of a burning flame. Hot air rises, dragging in colder air near the ground from the sides. Thus although the centre and top of a burning canopy may be above 1000 °C, temperatures may be expected to reduce with height to perhaps just a few hundred degrees, unless there is a marked radiation of heat downwards from burning fuel above ground level (such as in dense shrubbery). Even so, temperatures above 100 °C may persist for up to several minutes near the ground, especially if the 'burnout time' is long (see Box 5.1). The main key to surviving forest fires is to keep the heat of the flame away from living tissue. How this is done very much depends on whether you are considering plants or animals, and if looking at plants, what sort of fire is burning – whether it is a ground fire, a surface fire or a crown fire. Some plants will also continue via seeds if the parent is killed.

How plants survive a surface fire

The most common sort of fire is the surface fire that trickles or rushes through the dry surface litter of dead leaves and branches, and up into the shrub layer of the forest (see Chapter 3). These fires, although they produce high temperatures,

---

Box 5.1. How long does burning last?

The time that flames persist is called the **flame residence time**.

But once the flames are out, some smouldering combustion may continue; the total time of combustion (including smouldering) is called the **burnout time**.

For living material, the important factor is how long the surroundings stay hot. The time that temperatures stay above a threshold (say 100 °C) is called the **temperature residence time**.

can be survived. Two main mechanisms are used: survival by using thick bark and regrowing from new sprouts.

Thick bark

Thick bark makes a good insulator, protecting the sensitive living tissue inside a tree from the high temperatures of a passing surface fire. In fire ecology parlance, this is referred to as the tree being 'resistant' to fire. The depth to which lethal temperatures penetrate is related to fire intensity and the flame residence time (Bova & Dickinson 2005). The longer and more intense the fire, the more damage is caused. Generally, the thicker the bark, the longer the time taken for lethal temperatures to get through. Usually, this is a straight relationship such that twice the thickness of bark gives twice the protection, but it is not always so. If the bark starts to char or burn, as it does in some eucalypts and birches, then this obviously reduces the bark's effectiveness since it becomes progressively thinner. It has long been known that winter fires are generally not as damaging as summer fires and part of this may be because the bark starts cooler and more heat can be coped with before lethal temperatures reach the inside (but this is probably also to do with the physiologically dormant state of the tree as well). Dry bark is also normally a better insulator than bark with a higher moisture content since water is a poor thermal insulator. An exception can be found in the grey gum (*Eucalyptus cypellocarpa*) which has a thin, moist, non-flammable bark and yet responds to heat slowly (Vines 1968). When heated the outer layers evaporate water (steaming of the trunk is very noticeable) and so the temperature of the bark cannot rise above 100 °C until all the moisture has been evaporated.

Some barks are better insulators than others. The cork produced by the Mediterranean cork oak (*Quercus suber*) is a superb example. Air-filled cells in the cork create the same insulating effect that expanded polystyrene or a down-filled sleeping bag produce. In this case, of course, it is not to keep the heat in but to keep it out. This is why we make floor tiles from cork – it keeps the heat in our feet and thus feeling warm. Cork oaks live in flammable Mediterranean scrub that naturally burns every decade or so and they survive repeated fires by means of their insulating bark. The cork oak is not tolerant of shade and so benefits from the repeated fires that remove its competitor, including the thin-barked holm oak (*Quercus ilex*) with poorly insulating bark, which is normally restricted to areas where fire is uncommon. Survival is not surprisingly less where in trees that have had the bark removed commercially, leaving just a thin layer behind (Moreira *et al.* 2007).

Other trees with thick highly insulated bark can be quoted as equally good examples, including the Douglas fir (*Pseudotsuga menziesii*) of North America and some eucalypts (*Eucalyptus* spp.) from Australia; see also Box 5.2. The best example, shown in Fig. 5.1, is perhaps the giant sequoia of the Sierra Nevada Mountains of California.

(a)

(b)

Some trees have evolved a finely tuned relationship with fire. In the red syringa tree (*Burkea africana*), for example, a leguminous tree of the tropical savannahs of South Africa, bark thickness increases until the trunk is around 125 cm in diameter and then it almost ceases. This is also the time when seeds are first produced. It seems that once the bark reaches a set thickness, the tree is largely resistant to fire and the energy resources used to grow extra bark are switched into seed production (Wilson & Witkowski 2003).

Resprouting

Many trees in areas prone to surface fires do not have thick bark – think of poplars and birches. In their case, even a fairly gentle fire of a few hundred kW/m (one that could be jumped over) is likely to kill the living tissue (cambium and phloem) under the bark and lead to death above ground. However, many damaged in this way can resprout from below ground (Fig. 5.2). Soil is a superb insulator and even a hot surface fire does not usually produce lethal temperatures more than five or so centimetres down (see Chapter 3). So, trees that can sucker from the roots, such as aspen (*Populus tremuloides* in North America and *P. tremula* in Europe), and those that have stored buds buried below the lethal heat depth or can grow new buds afresh, as is often the case with birches (*Betula* species), will regrow from the ground. A study of aspen in western USA showed that up to 50% of the circumference of a stem could be charred without death, but charring more than 75% of the circumference resulted in death of the stem and extensive suckering (Brown &

Figure 5.1
The giant sequoia (*Sequoiadendron giganteum*) in Sequoia National Park, California amongst which are numbered the largest trees in the world. (a) These giants can grow to over 10 m in width and have a superb spongy, air-filled bark between 30 and 80 cm thick. The bark itself is virtually non-flammable and will at most burn by glowing combustion although, as can be seen, scorching by surface fires can be quite extensive. The growth of the roots immediately by the trunk adds another line of defence by raising the tree onto a slight mound; this can be seen here and also in Fig. 5.8b. Falling dead material thus tends to roll away from the tree keeping the most intense fires away from the base of the trunk. (b) Giant sequoias live in excess of 2000 years and naturally experience fire every 3–35 years – so the bark must work. But eventually the bark can be penetrated and the flammable wood is gradually removed by successive fires. This leaves scars inside the trunk (see Fig. 4.11 for an example) from which the local fire history can be traced.

(a)

(b)

(c)

DeByle 1987). Suckers of the aspen arose from up to 18 cm deep. In fire ecology terms, this ability to regrow from below ground is referred to as being able to 'endure' fire.

The best example must be those plants with lignotubers, a large woody swelling below or just above the ground holding a mass of buds. These lignotubers are found on many trees in dry tropical savannahs that are prone to frequent fires, including a number of South African proteas and especially in the 'mallee' eucalypts of western Australia. The largest lignotuber known was 10 m in width out of which sprouted 301 stems.

But it is not just hardwoods that can resprout. In areas of fairly high fire frequency a small number of pines have also evolved the ability to sprout from the stem or base of the trunk, at least when they are young – see Box 5.2. One of the best examples of a sprouting conifer comes from California: the coastal redwood (Fig. 5.2c). Being a tall redwood there is a large gap between the surface fuels and the

Figure 5.2
(a) New sprouts of mesquite shrubs (*Prosopis glandulosa*) from below ground after the stems above ground were killed 2 months previously by a fire in Texas. (b) Smaller herbaceous plants, such as the bunchberry (*Cornus canadensis*) and blueberry (*Vaccinium* species) shown here in eastern Canada are also frequent sprouters after fire. (c) These coastal redwoods (*Sequoia sempervirens*) in Muir Woods, California result from shoots growing up around an original tree that has long gone. The two trees on the right can be seen to be joined together.

Box 5.2. Adaptations to fire found in a selection of different pines (*Pinus* species). A dash indicates that the adaptation is not found in that species; the more pluses, the more important the adaptation in that species. Based on Richardson (1998) with common names (all followed by 'pine') and geographical origins taken from Welch & Haddow (1993). The different adaptations are described in the text

| Pine species | | Origin | Grass stage | Resprouting | Serotiny | Thick bark | Self-pruning |
Scientific name	Common name						
P. attenuata	Knobcone	USA/Mexico	–	–	++++	–	–
P. banksiana	Jack	NE America	–	–	++++	–	–
P. brutia	Calabrian	Turkey etc.	–	–	++	+	+
P. canariensis	Canary	Canary Is.	–	++++	–	+++(+)	–
P. contorta	Lodgepole	Rocky Mtns	–	–	++++/–	–	–
P. coulteri	Big-cone	SW USA	–	–	–/+	++	–
P. devoniana	Michoacan	Mexico	++++	–	–	++	++
P. echinata	Shortleaf	USA	–	++++	–	+	+
P. greggii	Gregg	Mexico	–	–	++++	+	–
P. halepensis	Aleppo	Medit. Europe	–	–	++/–	+++	+
P. jaliscana	Jalisco	Mexico	–	–	++++	+++	+++
P. jeffreyi	Jeffrey	USA	–	–	–	++++	++++
P. leiophylla	Smooth-leaf	Mexico	–	++++	–/+	+++	+++
P. merkusii	Sumatran	SE Asia	++++	–	–	?	?
P. montezumae	Montezuma	C. America	++++	–	–	+++	+++
P. muricata	Bishop	W. USA	–	–	+++	+++	+
P. oocarpa	Egg-cone	Peru	–	++++	++++/–	+	+
P. palustris	Longleaf	S-E. USA	++++	–	–	++	++++
P. ponderosa	Ponderosa	N America	–	–	–	++++	++++
P. pseudostrobus	Smoothbark White	C America	–	–	–	+++	++++
P. serotina	Pond	USA	–	++++	++++	+	–
P. torreyana	Torrey	California	–	–	–/+	+	+

canopy and fires are resisted by a spongy air-filled bark. This bark is not as thick as in the giant sequoia described above so eventually fire can get through to the flammable wood. Repeated fires along with rot will eventually damage the tree so much that it breaks. A crown fire might also kill the canopy. Either way, by repeated burning of the trunk or a crown fire, the top of the tree can die. Fortunately for the coastal redwood, it is capable of sprouting from dormant buds if the top is killed, producing a circle of fine new shoots around its circumference. This regrowth is especially vigorous on young trees, with shoots reaching over 2 m in the first year. Top-killed trees thus produce new sprouts, which in turn grow into mature trees with their own roots, sometimes still joined together at the base into a full or partial circle (as in Fig. 5.2c). As these trees are themselves eventually killed by repeated fires, they in turn will produce new sprouts and, since the ones on the outside are likely to get most light and survive, the circle grows bigger like the ripples from a stone dropped into water. Inevitably, trees

(a)

(b)

around the circle periodically die and the remaining arcs blend into the surrounding forest and the process begins again.

Trees with thick bark on the trunk and main branches may also survive fire by producing new growth from surviving buds above ground if the foliage is killed by rising heat. In such cases trees can produce a 'fire column' – a tall tree with short feathery foliage which eventually grows back into proper branches (Fig. 5.3). Similarly, the cork oak of the Mediterranean normally survives fire using its thick bark but if the foliage is burnt in a very intense fire, it resprouts from the bigger boughs, producing a wider fire column. Again, eucalypts provide some of the best examples, with resprouting possible from the trunk along to the thinnest branch. It has been suggested that the shape of a tree may help in protecting buds from high temperatures. For example, in South Africa, the inverted conical shape of the laurel protea (*Protea laurifolia*) and the silver tree (*Leucadendron argenteum*) may help protect buds by deflecting away hot air currents rising from the fire below (the shape may, however, be a consequence of lower branches being killed by previous fires rather than an evolved strategy to deflect heat).

Box 5.2 may give the impression that trees can either sprout or not: all or nothing. In reality, sprouting ability can vary tremendously between species. In Japan, Masaka and colleagues (2000) observed that the Japanese paper-bark birch (*Betula platyphylla* var. *japonica*) produced up to 296 sprouts per tree, especially on larger trees (more buds stored) and more seriously damaged trees (diverting more resources from resprouting/regowth in the canopy). On the other hand, they found

Figure 5.3
Fire column. (a) Epicormic resprouting after crown scorching by fire from preformed buds on a canary pine (*Pinus canariensis*) from the Canary Islands. From Climent *et al.* (2004) with kind permission of Springer Science + Business Media. (b) New sprouts from a stringy bark eucalypt (*Eucalyptus* species) in Victoria, Australia, three weeks after an intense fire. Photograph by Mike Benefield.

(a)

(b)

Figure 5.4
(a) Grass trees in western Australia showing the persistent leaves that cluster round the growing tip, helping protect it from fire. (b) A burnt grass tree with new growth above the burnt stem. This plant is growing in a large greenhouse at the National Botanic Garden of Wales in the UK and has been burnt with a blow torch to simulate a wildfire. Photograph (a) by Richard Hobbs.

that Mongolian oak (*Quercus mongolica* var. *grosseserrata*), relies more on its thicker bark and regrowth of the canopy, and produced a maximum of 34 sprouts per tree even when severely damaged. Sprouting also varies with age of the tree. A study of red oak (*Quercus rubra*) in the USA by Johnson (1975) found that sprouting declined from 56% of 20 cm diameter trees to 33% of 50 cm trees to no sprouting in 66 cm diameter trees.

Growing points can be protected by plant parts other than just bark. For example, some cycads (primitive distant relatives of conifers) and the grass trees (*Xanthorrhoea* species) of Australia have the growing point protected by the dense rigid leaf bases that cluster around the growing tip and are glued together by gum exuded by the trees (Fig. 5.4). Many grasses also benefit from masses of dead foliage protecting the growing points.

A number of pines in warm, infertile areas around the world that have frequent but low-intensity fires have pines that have a 'grass stage' – see Box 5.2. The most studied is the longleaf pine (*Pinus palustris*) shown in Fig. 5.5 growing on the sandy Pine Barrens of south-east USA. Adult pines are safe from the normal low-intensity fires because of their thick bark but young saplings, with their thin bark and canopy readily reached by the flames, are very vulnerable. So, for the first 5–20 years the pines grow very little above ground, and the vulnerable growing tip is protected by dense fire-resistant needles. During this time the root system grows enormously and food is stored. Finally, the seedling bolts upwards, growing as much as 1.5 m in 3 years. During this bolting stage they are vulnerable to fire but quickly develop a thick bark and self-prune dead branches to remove 'ladder fuels' (see Chapter 4)

(a)

(b)

and reduce the risk of lethal crown fires. The strategy is to spend as little time in the danger zone as possible.

It is not just trees that can sprout from below ground; a similar ability is found in a wide range of shrubs and herbaceous plants. In eastern Canada, Flinn & Wein (1977) found that the roots of many small woodland herbs, such as wintergreen (*Gaultheria procumbens*) and starflower (*Trientalis borealis*), were in the top few centimetres of soil, whereas underground stems of plants like bracken (*Pteridium aquilinum*) were up to 12 cm down. Since different species tend to grow at different depths below ground, it is to be expected that fires of increasing intensity (with a deeper penetrating lethal heat pulse) are likely to result in very different post-fire communities.

Figure 5.5

(a) The grass stage of longleaf pine (*Pinus palustris*) in Florida. The seedlings stay short, putting their resources into growing roots before growing rapidly upwards through the sapling stage when they are at risk from damage by fire. (b) Two seedlings that were caught and killed at the vulnerable stage.

Fire stimulation of flowering

A number of plants that endure fire and sprout afterwards take additional advantage of the good growing conditions after a fire by producing more flowers, and thus more seed. In Australia the most spectacular example are the grass trees (Fig. 5.4) where burning stimulates the formation of flowering shoots. This strategy is especially common in the fynbos shrublands of South Africa with spectacular post-fire displays. Such plants are mostly non-woody monocotyledonous plants (one of the main divisions of the flowering plants), including the Australian grass trees and a range of orchids, grasses and members of the iris, lily and daffodil families (Iridaceae, Liliaceae and Amaryllidaceae, respectively). Many of these plants are 'geophytes' with some sort of bulb or corm underground from which they can regrow, and which may remain dormant between fires. Flowering can be very rapid after a fire; the fire lily (*Cyrtanthus contractus*) of the Eastern Cape of South Africa flowers within 14 days of a fire. This enables it to make the most of pollinators and grow new seed before other plants have begun to recover from the fire and start competing for light and water. Some species may flower only after a fire and are thus wholly dependent upon fire for the next generation.

What causes the extra flowering? It may be extra nutrients from the ash, the greater fluctuation in the daily temperature regime or moisture, decreased competition or even increased light (Brewer *et al.* 2009). In the case of the grass trees, pruning by the fire appears to play a part since experiments have shown that clipping gives roughly the same result.

Ground fires and plant survival

The one type of fire that is difficult for a plant to survive is a ground fire which burns through the organic matter layer that builds up on top of the mineral soil, especially in cold, moist climates. As described in Chapter 3, this organic layer burns slowly by glowing combustion with little flame (in the same way that a cigarette burns). Since most of the roots holding the tree up and keeping it supplied with water and nutrients are in this organic layer, they are slowly but surely consumed. Fallen trees often look completely untouched (Fig. 3.5) but the lack of roots means death. Even if the roots are not directly burnt, the heat load from the glowing combustion can kill the entire base of the tree. Certainly, smouldering fires with temperatures above 300 °C for 2–4 hours are not unusual, producing enough energy to penetrate even thick bark, readily killing the phloem and cambium. Seeds of herbaceous plants and of trees are also mostly buried in the same layer and so can also be consumed (unless a very wet layer of organic matter remains unburnt and unheated below the fire). New trees thus normally have to reinvade from seeds blown in from outside the fire area. There is little biological adaptation to this sort of fire!

How plants cope with a crown fire

The really big challenge to trees comes with crown fires. As described in the last two chapters, a crown fire is where the fire manages to bridge the gap between the surface fuels and the flammable tree canopy to produce an inferno releasing 30 000–100 000 kW from every metre-wide stretch of fire as it burns forward.

How can trees possibly survive this? In northern hemisphere forests dominated by pines the simple answer is that they usually do not. Some thick-barked trees might resprout from surviving buds as described above, but the majority of trees are incapable of withstanding the huge energy load of a roaring crown fire. In forests dominated by hardwoods (i.e. not conifers) a larger number of plants will survive by sprouting from below ground (as is the case in places like Australia – Clarke & Dorji 2008) but gaps will be left where plants are killed.

The key to replacing the dead plants is that although the individual plant might die, there are mechanisms that will ensure the survival of the species on that site. Like a phoenix, they rise from the ashes. The most important mechanism is seed that is stored either in the soil or stored in the canopy. These are the 'fire evaders'.

Seed storage in the canopy: serotiny

Many pines and hardwood trees that are regularly involved in frequent, intense crown fires have their seeds stored in fire-proof cones or fruits which open when burnt or heated. This is referred to as 'serotiny' (but see Box 5.3). At least part of each year's seed crop is stored in the canopy, building up a reserve of seeds ready to be released after fire.

At least 500 different woody trees and shrubs around the world are serotinous (Boxes 5.2 and 5.4) with perhaps another 200 still to be added. The majority are found in the southern hemisphere, where serotiny is most common in evergreen hardwoods (such as the proteas, banksias and eucalypts of the Cape area of South Africa and particularly the shrublands (bush) of south-west Australia. Of all these, the Australian bush has the largest set of serotinous species. In the northern hemisphere, serotiny is most common in the pines (particularly in North America)

Box 5.3. The pedant's guide to serotiny

Serotiny comes from the Latin *serotinus* 'late in occurrence', i.e. seeds are stored in fruits and cones and their shedding is delayed.

The term **bradyspory** (applied to a plant that disperses its seeds slowly; from *brady*, Greek for slow, and *spory*, Latin for spores, or, in effect, seeds) or, more commonly now, **bradychory** are slightly wider terms, encompassing all cases of delayed dispersal (such as seasonal patterns), not just fruits and cones which stay closed.

Strictly speaking, the term serotiny refers to the holding of seeds rather than the release of them by fire and it is possible for serotinous seeds to be released by triggers other than fire, such as wetting and drying cycles (xeriscence). In which case, we should strictly refer to the release of seeds by fire as **pyriscence** (*pyr*, Greek for fire and *hiscere*, Latin for gape – what the fruit/cone does when it releases its seed).

Usually, however, serotiny is used as an overall term to mean the whole process of storing seeds in the canopy and their release by fire.

Box 5.4. A selection of serotinous trees and shrubs that store their seed in fire-proof fruits, which are released following fire. See also Box 5.2

Tree		Type of tree	Origin	Type of fire-proof fruit
Scientific name	Common name			
Actinostrobus species	Cypress-pines	Conifer (Cupressaceae)	W Australia	Round cones
Banksia species	Banksia	Hardwood (Proteaceae)	Australia	Dry woody fruits embedded in persistent woody bracts (modified leaves)
Callitris species	Cypress-pines	Conifer (Cupressaceae)	Australia	Round cones
Cupressus forbesii	Tecate cypress	Conifer (Cupressaceae)	W USA	Round cones
Erica sessiliflora	Heath	Hardwood (Ericaceae)	South Africa	Dry capsules
Leucadendron argenteum	Silver tree	Hardwood (Proteaceae)	South Africa	Cones
Pinus attenuata	Knobcone pine	Conifer (Pinaceae)	USA/Mexico	Cones
P. banksiana	Jack pine	Conifer (Pinaceae)	NE America	Cones
P. contorta	Lodgepole pine	Conifer (Pinaceae)	Rocky Mtns	Cones
P. greggii	Gregg pine	Conifer (Pinaceae)	Mexico	Cones
P. halapensis	Aleppo pine	Conifer (Pinaceae)	Medit. Europe	Cones
P. jaliscana	Jalisco pine	Conifer (Pinaceae)	Mexico	Cones
P. muricata	Bishop pine	Conifer (Pinaceae)	W USA	Cones
P. oocarpa	Egg-cone pine	Conifer (Pinaceae)	Peru	Cones
P. pinaster	Maritime pine	Conifer (Pinaceae)	Medit. Europe	Cones
P. radiata	Monterey pine	Conifer (Pinaceae)	California	Cones
P. serotina	Pond pine	Conifer (Pinaceae)	USA	Cones
Protea species	Protea	Hardwood (Proteaceae)	South Africa	Dry woody fruits embedded in persistent woody bracts (modified leaves)
Tetraclinis species	Arartree	Conifer (Cupressaceae)	Medit. Europe	Round cones
Widdringtonia species	African cypress	Conifer (Cupressaceae)	South Africa	Round cones

where the seeds are stored in sealed cones (Fig. 5.6). In all these plants the seeds are held in a range of fire-proof structures from the small capsules of the eucalypts to the massive woody 'infructescences' (cones) of the banksias (really a mass of modified, woody leaves fused together) bearing a number of woody fruits (follicles), each containing at most two seeds.

Seeds may be stored in the canopy for decades. In banksias, for example, seeds remain viable in the canopy for 10–15 years and in conifers this may stretch to 20–30 years or even longer (in rare cases to over a century). Come a fire, seed release is usually triggered by heat. In conifers, the scales of the cone are sealed shut with resins that prevent the cones opening as normal. But a fire hot enough to kill the mother tree will melt these resins (it generally needs temperatures at the cone of at least 45–60 °C) and the cones will then open normally a few days after the fire. The fruits of banksias open when heat ruptures the follicle valves; this requires temperatures of 100–300 °C for at least 2 minutes. This is aided by serotinous species of banksia keeping the dead flowers on the fruits which burn adding extra

(a)

(b)

Figure 5.6
(a) Serotinous cones on burnt lodgepole pine (*Pinus contorta*) in the Canadian Rocky Mountains. The needles are gone and the masses of cones can be seen along the branches. These will have been heated sufficiently that within a few days the scales will open and the accumulated seed will drop to the ground, starting the next generation. (b) The jack pine (*Pinus banksiana*) cone on the right was heated in an oven a few hours earlier to 60 °C for 2 minutes and has subsequently fully opened while the unheated cone on the left stays resolutely closed.

heat. Eucalypts shed seed when the capsules or the branch they are on (or even the tree) are killed by fire and the capsules then dry out, shedding seeds hours to weeks after the fire.

Crown fires normally provide the sort of heat needed to open the cones and fruits. However, serotinous structures can sometimes be opened by a surface fire. For example, in the Australian mountain ash (*Eucalyptus regnans*) a surface fire, which does not scorch the canopy, may girdle and kill the stem, causing the capsules to dry out and shed their seeds. This is especially rapid (in a matter of weeks) if the litter and humus is burnt, and the roots are killed.

These serotinous seed containers obviously have to be fairly heat-proof to prevent the seeds dying in a fire. Fraver (1992) heated the small cones of pitch pine (*P. rigida*), 2.5–3.5 cm wide, for 3 minutes either in an oven or over a flame and found that an internal temperature of 100 °C was not reached until there was an external temperature of around 420–500 °C. Moya *et al.* (2008) looked at serotinous and non-serotinous cones of the Mediterranean aleppo pine (*Pinus halepensis*) and found that temperatures inside serotinous cones reached more than 70 °C in simulated fires – above the normal lethal limit of living tissue. The temperature inside non-serotinous cones was, on average, only 2–17 °C lower than in serotinous cones, depending upon the intensity and duration of the fire. Fortunately, dormant, dry seeds can usually withstand temperatures up to around 100 °C for 5 minutes. Moreover, dry seeds of jack pine (*Pinus banksiana*) have been exposed to 370 °C for 5–20 seconds and even 430 °C for 5–10 seconds without deleterious effects. So the role of serotinous cones does not appear primarily to protect the seeds from heat, but to store seeds to be released after fire. Indeed, small cones of pines such as the pitch pine of the New Jersey Pine Barrens can be completely charred all the way through. In some trees the heat transmitted through the cone appears to be beneficial to the seeds. Goubitz and colleagues (2003) even found that seeds from serotinous cones of the aleppo pine germinated better than seeds from non-serotinous cones once they had been exposed to 100 °C for 5 minutes and laid on a seedbed at pH 10, equivalent to a seedbed with a coating of ash.

Serotinous pine cones, as well as being sealed shut, generally are more compact and rigid with thicker and bigger scales (Moya *et al.* 2008). The small fruits of eucalyptus may be less heat-proof simply because they have less woody material in them. However, clumping of fruits together may provide greater fire protection. Ashton (1986) records that after a fire, the outermost capsules of the messmate eucalyptus (*Eucalyptus obliqua*) were charred while those in the middle of a bunch were only lightly charred or even still green.

Usually within a few days of the fire when the ground is cool (but sometimes over several months), the majority of seeds fall onto the ground. A number of pines will keep some seeds at the base of the cone which are subsequently released little by little over the next few years (perhaps as an insurance policy?) but most seedlings get started in the first year after a fire, and certainly within the first few years. The seeds usually do well because they are on a most favourable seedbed that provides for their needs:

- Warmth – the blackened surface absorbs heat and thus speeds germination. On the negative side, the very thin top layer of the blackened surface can reach temperatures in excess of 80 °C, but there is so little heat stored that seeds are unlikely to be harmed although seedlings can be cooked and killed by that thin hot layer at soil level.

- Plenty of nutrients from the ash – thick layers of ash from wood can, however, be very rich in hydroxide and bicarbonates, and thus extremely alkaline (which is why wood ashes were used in soap making). When wet this 'soup' can rapidly kill seeds (Thomas & Wein 1990). But thin layers, or deeper layers that have been leached by rain may be beneficial to seed germination (see below under soil seed banks) and help new seedlings to grow vigorously.

- Little shade from above – many serotinous species are shade-intolerant and will grow very slowly or not at all in the deep shade of their intact parents.

- Little litter – layers of leaf litter dry rapidly after rain, and young seedlings are prone to die if the litter dries before their roots reach the more constant moisture lower down; removal by fire solves that problem.

- Little immediate competition – removal of the canopies of herbs and shrubs results in reduced competition for light above ground and water below ground. Sprouting plants will rapidly reassert themselves but young seedlings get a valuable breathing space in their first few vulnerable weeks.

- Charcoal – there is evidence that charcoal improves germination and seedling growth of some trees by improving their nutrient uptake, perhaps by stimulating soil microbes and thereby increasing litter breakdown, and by absorbing toxins from the soil (Wardle *et al.* 1998, Hille & den Ouden 2005).

Provided water is available, the next generation of serotinous plants will rapidly dominate the site. In the case of jack pine in the boreal forest of Canada, as many as 4 million seeds per hectare can fall after the fire, resulting in 'dog's hair regeneration' – seedlings as thick as the hair on a dog's back (Fig. 5.7). If we planted seeds that dense it would mean one seed every 5 cm on a square grid.

As well as ensuring a good start to the next generation, the intense crown fire that releases the serotinous seeds also kills the fire-intolerant competitors that would otherwise take over the stand. In the case of jack pine forests, these would be the shade-tolerant spruces and firs. Thus the fire helps to ensure that jack pine will beget another almost pure jack pine stand.

It is thus in the evolutionary interests of a serotinous species that fires that pass through their stands are good intense ones, and it appears that most serotinous trees have evolved to become very flammable (this has been hotly argued and it may be the

Figure 5.7
Jack pine (*Pinus banksiana*) regeneration 7 years after the parent trees were killed by a fire in the Northwest Territories, Canada. When the photograph was taken there were 2.4 million seedlings per hectare, so close together that it was difficult to put a foot down between the stems. The dead parent trees typically stand for 25 years or more before a wind storm finally brings the majority down like dominoes.

other way round; plants that are flammable have evolved reproductive strategies to cope and take advantage of it – see Bond & Midgley (1995) for an evolutionary argument). Regardless of its origin, anyone who has seen news footage of the bushfires of south-west Australia over the past decade will not find the idea too surprising that flammability and serotiny go hand in hand. One of the main traits that aids flammability is the retention of 'ladder fuels' that help carry a gentle surface fire up into the canopy to produce the right sort of stand-replacing fire that allows the next generation to start. The ladder fuels are mostly formed by dead (and hence flammable) branches down to ground level, often encrusted with flammable lichens, that are not shed or 'self-pruned' (Fig. 5.8). Most serotinous species also have flammable bark, such as in the pines and many of the eucalypts, due to high levels of resins or oils. Other trees, such as the stringybark eucalypts, end up with dead lengths of flammable bark draped over the branches to the ground.

Serotiny is of most benefit in areas that have fairly frequent high-intensity fires. If fires become too frequent, there will be insufficient time to accumulate the next crop of seeds and trees with other strategies, such as surviving the fire by a thick bark or resprouting, will do better. If fires are not frequent enough or not intense enough, seeds will not be released often enough and non-serotinous trees will have the advantage. Since serotiny appears to be under fairly simple genetic control, natural selection can quickly operate to change the degree of serotiny in a species. Generally as the frequency of fires lethal to the parents becomes less, the amount of serotiny declines, so populations of the same species in different areas may have different amounts of serotiny along a gradient of all to none, depending upon the local fire regime (see Box 5.5). Moreover, in any one area,

(a)

(b)

individual trees with lower serotiny increase with the time since the last lethal fire, since trees with lower serotiny will be able to produce more seedlings in the longer gaps between fires.

Trees with lower serotiny can be 'mixed' trees, carrying both serotinous and non-serotinous cones; in the case of the European aleppo pine (*Pinus halepensis*) and probably others as well, the serotinous cones are concentrated at the top of the tree. This mixture allows a trickle of seed to fall each year and so allows some regeneration in the absence of fire. The ability to produce mixed trees does depend upon the species: for example, jack pine is more likely to form mixed trees than the closely related lodgepole pine or pitch pine (*Pinus rigida*). Another way of being partly serotinous is to vary how long fruits stay closed. Some proteas in South Africa and some North American pines are 'weakly' serotinous such that the fruits stay closed for just a few years and then open even without fire. For example, the Monterey pine (*Pinus radiata*), which grows naturally on the Monterey Peninsula of California and has been planted extensively around the world, keeps huge numbers of cones in the canopy,

Figure 5.8
Flammable ladder fuels. (a) The two jack pine trees (photographed in Ontario, Canada) show the two main types of ladder fuels that help lift a surface fire into the canopy of the trees. The tree in the foreground has retained dead branches and the one in the background demonstrates that the bark itself is flammable. (b) A giant sequoia (*Sequoiadendron giganteum*) forest in the Sierra Nevada Mountains of California that has been unburnt for a number of years. Flammable dead branches and other fuel have accumulated at the base of the tree, and the fire-intolerant white fir (*Abies concolor*) has started to invade. The fir has plenty of ladder fuels in the form of dead branches and the yellow lichen that adorns them (this is wolf lichen, a *Letharia* species). When a fire eventually starts it will be intense and readily climb the ladder fuels to potentially damage the crown of the giant sequoia which does not usually meet fire.

Box 5.5. Variation in serotiny

The lodgepole pine of the boreal forest (*Pinus contorta* var. *latifolia*, left) meets fire frequently – every 100–150 years on average – and is usually highly serotinous. Stands are uniform in age, fairly short-lived and regenerate from seeds dropped almost exclusively after the fire. If fire is removed, shade-tolerant (and fire-intolerant) spruces and firs gradually replace the lodgepole pine.

In mountains to the south (the Cascades of Oregon and Washington, the Sierra Nevada Mountains of California, for example) the lodgepole pine (right) is sufficiently different to be called a different variety (var. *murrayana*); here it is a subalpine tree which meets fire very infrequently, and it is uniformly non-serotinous (right). It lives for more than 600 years, forming uneven-aged stands with seedlings establishing in gaps created by individual trees dying. Groups of the pine are probably not taken over by more shade-tolerant trees due to the harsh conditions caused by the thin soils.

festooning even the larger branches. These are serotinous but only weakly so; the cones remain closed at maturity but after a few years open as normal without a fire. This may be due to successive warm summers loosening the resin that bonds the scales closed (serotinous pine cones are known to open on hot sunny days, especially if lying on a hot litter surface). Eucalypts, on the other hand, shed their seed regularly and slowly with a more major shedding after fire. The Australian mountain ash (*Eucalyptus regnans* – the tallest hardwood tree in the world) is known to shed seeds throughout the year but especially after a hot dry summer or a fire, when 50–75% of the year's seed shed may occur.

Weak or partial serotiny may not just be a strategy for aiding seedling establishment between fires. Cowling and co-workers (1987) found that in four *Banksia* species in Australia, seedling recruitment was almost completely confined to the immediate post-fire period. Seeds dribbled out of some cones in the autumn and germinated but the seedlings largely died. Similar stories have been found of weakly serotinous species in South Africa as well. It may well be that the costs of maintaining a large serotinous seed store are too high (all the cones/fruits consume energy to keep them alive), and the weak serotiny is a way of getting rid of the oldest seeds while still keeping a good store on the plant (see Midgley 2002 for a more detailed argument).

There has been a good deal of debate as to the advantages of storing seed in the canopy rather than the soil. Storage in the soil must be cheaper because there are no big heat-proof cones or fruits to grow (which can take more resources to grow than the contained seeds). The advantages of serotiny seem to boil down to more certain protection from fire, and also defence against herbivores by keeping the large, easily found seeds that most trees have off the ground (Lamont *et al.* 1991 gives more detail).

Seed storage in the soil

Many plants shed seeds as normal but these do not germinate immediately and are incorporated into the litter and humus to form a soil seed bank. Some plants more readily form a soil seed bank than others. These include many herbaceous members of the pea family (Fabaceae), woody shrubs in the rockrose family (Cistaceae) and a number of southern hemisphere plants from families such as the Proteaceae and Mimosaceae (e.g. Edwards & Whelan 1995). Some seeds remain in the soil just over an inhospitable time of year (this includes the many annual species typical of the African savannahs) while others may be stored for a few years (e.g. Buhk & Hensen 2008) up to a number of decades (Keeley & Fotheringham 2000). Trees are less good at forming soil seed banks (probably because their seeds tend to be large and so easily found by mice etc.) but include a number of the cherries with their hard stone around the seed and acacias with hard seed coats. In South Africa, as many as 125–250 million seeds per hectare of acacia species with small seeds have been found in the top 10 cm of the soil (Milton & Hall 1981), although it should be noted that these authors also report that 92% of the soil seed bank can disappear after a single fire. Not all species in a fire-prone area store large numbers of seed in the soil. A few pines, such as jack pine, store seeds from one season to the next but this is really just delayed germination rather than a long-term soil seed bank (Thomas & Wein 1985). One pine comes a lot nearer to having a soil seed bank: the whitebark pine (*Pinus albicaulis*), of western North American mountains, has seeds that are dispersed and buried by the Clark's Nutcracker – the birds use these cached seeds for their winter food supply, but some are forgotten or are unused and are left ready planted. These cached seeds may not germinate for up to 2 years while the small embryo in each seed slowly matures (Tomback *et al.* 2001). Although whitebark pine seedlings do appear on burnt areas this soil storage strategy probably isn't really fire-related; the main advantage to the tree appears to be in ironing out of infrequent seed production years so that there are always some seeds available to take advantage of wet periods.

Quick burial of even small seeds will reduce the risk of being eaten, and the risk of being cooked by a fire. This is aided in many Mediterranean-type plants (especially in Australia) by seeds possessing an 'elaiosome', a small oily, protein-rich body which is eaten by ants. The ants carry the seeds to safety underground, chew off the elaiosome, leaving the seed unharmed and nicely buried out of harm's way. Larger seeds may also be buried by larger animals such as rodents. For example, the large seeds of the big cone pine (*Pinus coulteri*) are too heavy to be blown far by wind and so tend to fall from the serotinous cones close to the parent tree. But they are then secondarily dispersed by rodents; seeds are taken and buried safely below ground often more than 10 m from the parent tree (Borchert *et al.* 2003). Some of these seeds will be eaten by the rodents but some will remain untouched and able to germinate.

Storage of seeds in the soil is a particularly useful strategy in places where the fire frequency is longer than the normal lifespan of the plant – serotinous species would die and rot with their seeds. Most species do best in open seedbeds (for the same reasons seen in serotinous species) and so do best to germinate after a fire. The biggest drawback of a soil seed bank is that the seeds are vulnerable to high-intensity fires that heat the soil and kill the seeds, or high-severity fires that burn away the organic matter, consuming the seeds.

What makes buried seeds germinate at the right time?

The major problem for a buried seed at some depth in the soil is the need to receive a stimulus when a fire has burnt across the surface. The seeds are dormant and need some sort of trigger produced by the fire to break their dormancy and stimulate them to germinate.

The first potential trigger is direct stimulation by the heat of the fire; this is found, for example, in around 15% of the fire-prone plants in south-west Australia (Auld & Denham 2006) and is a common strategy in members of the buckthorn and rockrose families (Rhamnaceae and Cistataceae) and a number of others, but is also found in such plants as the aleppo pine (*Pinus halepensis*) in the Mediterranean and even the Scots pine (*P. sylvestris*) of Europe (Núñez & Calvo 2000). However, response to a direct heat pulse is most common in 'hard-coated' seeds such as those commonly produced by legumes. The seed coat is hard and thick and prevents germination by keeping oxygen and especially water away from the seed inside. The heat pulse penetrating into the ground leads to rapid expansion of the seed coat, causing it to crack, allowing in oxygen and water and thus triggering germination. This can be a subtle process in that too little heat and the seed will not germinate; too much heat and the seed will be killed. Fortunately, as seen above, dormant, dry seeds are quite resistant to heat. Buried seeds have been known to survive 200 °C for 5 minutes (Gashaw & Michelsen 2002), somewhat more than the 80–120 °C for a matter of minutes that is the usual requirement for triggering germination. In Australia smaller seeds have been shown to withstand a greater heat shock and still germinate (Hanley *et al.* 2003) and this may well be true elsewhere. Thus while Bond *et al.* (1999) predicted that larger-seeded species should do better after an intense fire because their bigger seedlings can emerge from greater depths in the soil, more smaller-seeded species near the surface may survive the high heat pulse to germinate and so complicate the picture. Often seeds from within a species will vary tremendously in the amount of heat they need to trigger germination; this allows some seeds with a low threshold to germinate between fires and ensures that some with a very high threshold will remain ungerminated after a fire, keeping a reservoir (a residual soil seed bank) in case those that germinate after one fire all subsequently die. Having said this, a potential problem of management fires (usually burnt when conditions will produce a low-intensity fire and so be easily controllable) is that the heat pulse will be insufficient to trigger germination. This has certainly been seen to be a problem in cool autumn prescribed fires in south-east Australia (Penman & Towerton 2008).

Germination can occur if the seed coat is broken by other agents instead, such as repeated heating/cooling, drying/wetting, decay or passing through the gut of an animal. In species with hard-coated seeds these triggers can produce a trickle of germination (a useful strategy for taking advantage of small gaps as they appear) with the bulk occurring after a fire. Hard-coated species also produce a proportion of soft-coated seeds that germinate immediately.

Heat can also trigger germination in a more subtle way. Once the vegetation cover has been removed by a fire, day and night temperatures will be higher and lower, respectively, as the sun beats in during the day, and heat radiates out during the night. Some seeds, including those of many heathers and other shrub/bush plants, can detect this and use it as the trigger to germinate.

The chemical products of fire are also implicated in stimulating germination. The most famous trigger, perhaps, is smoke. In the shrublands of California, South Africa and Australia, the seeds of over 170 different plants are known to germinate in response to smoke (see Van Staden *et al.* 2000 for further details). Indeed, a number of Australian plants that were 'totally unamenable to conventional propagation from seed' (Dixon *et al.* 1995) germinate very well once they are given extracts of smoke. But smoke is not beneficial to every plant;

Reyes & Trabaud (2009) found that the germination of some European Mediterranean woody and herbaceous plants was improved but in others it was inhibited or not affected one way or the other. Similar results have been found in Australia (Ne'eman *et al.* 2009). This works either by the seeds being directly exposed to the smoke, or by rain dissolving the smoke from the atmosphere or from the soil surface where it has condensed. Thus, seeds deep in the soil – deeper than would be reached by a heat pulse – can be stimulated to germinate. The active ingredient of smoke appears to be a group of compounds called butenolides (Flematti *et al.* 2004). Smoke extracts are commercially available to help germinate these difficult seeds.

Germination of some plants has also been seen to be affected by a range of fire products:

- **ethylene and ash:** ethylene is released by wet ash up to levels around 0.5 parts per million (ppm) – germination of *Rhus coriaria* on Mount Carmel, Israel is certainly increased by ethylene levels a tenth of this (0.05 ppm) (Ne'eman *et al.* 1999);

- **charred wood:** the absorption of germination-enhancing chemicals onto charcoal, which are then washed off into the soil, may explain why charcoal is involved.

Other seeds may be dormant due to inhibiting chemicals (allelochemicals) in the soil produced by the plants already there (this is common in the dry climates of such places as California and South Africa). Seeds are 'released' to germinate when these chemicals are denatured by the heat of a fire.

Some plants may be triggered to germinate by a combination of the above factors. For example, dormancy in *Grevillea* species in Australia is known to be broken by a combination of smoke and heat shock (Auld & Denham 2006).

Sneaking past – invasion after a fire

A large number of 'invaders' – plants with small, easily dispersed seeds that are not fire-adapted – can take advantage of the open, optimal conditions after a fire. The plant that usually comes immediately to mind in the northern hemisphere is fireweed (more romantically called rosebay in Britain – *Chamerion angustifolium*) which can form almost pure sheets of purple after a fire, as in Fig. 5.9. It rapidly invades using a prolific seed production but to a certain extent it is adapted to cope with fire as an adult plant in that it can regrow from underground, horizontal stems (termed rhizomes). Trees can also be invaders, including the common pioneers of bare areas such as birches and willows. Some pines also come into this category, such as the eastern white pine (sometimes called the Weymouth pine, *Pinus strobus*) of eastern North America. It is long-lived, not particularly fire-adapted (although the bark will resist low-intensity fires) but it

will readily seed into burnt areas from surviving trees. Also included here is European Scots pine (*P. sylvestris*): it is a generalist that is not specifically adapted to fire but on fertile areas fire temporarily knocks back competitors allowing Scots pine to establish in the open. On more marginal sites it will survive very well without fire.

The final group of plants that need to be considered are the late arrivals; the 'avoiders'. These are often shade-tolerant plants that invade dense shady forests and are readily ousted by fire. They include such trees as many of the spruces (*Picea*) and firs (*Abies*) of the northern hemisphere. Also included here are the species that avoid fire by living in virtually fire-proof rocky or damp areas, but which can invade open burnt areas. Included here are natural populations of the European olive (*Olea europaea*) and the South African wild olive (*Olea africana*) and, interestingly, African populations of the heather *Erica depressa*; interesting because most *Erica* species elsewhere in the world readily cope with fire by resprouting.

Figure 5.9
Fireweed (*Chamerion angustifolium*) forming a dense carpet after a fire in the Canadian Rocky Mountains. It is not particularly adapted to fire but is a good invader of open spaces; fires just happen to be a good creator of these open spaces, hence the association and name.

Combining these strategies

In summary, fire-adapted plants exhibit two main sets of traits, either fire-surviving (using thick bark and tall trunks with no ladder fuels) or fire-embracing (with little investment in survival, enhanced flammability, using serotinous cones or a soil seed bank). But plants may not fall into just one camp or the other; bets are hedged. For example, pitch pine (*Pinus rigida*), native to eastern North America, can be serotinous or not but young trees (up to the age of 80) also have the ability to sprout from the trunk, its base and from branches. Other pine examples with several strategies can be picked out from Box 5.2. Away from conifers, the Bot River protea (*Protea compacta*) and the heather *Erica sessiliflora* (the only serotinous heather), both of South Africa, are serotinous but seeds also show enhanced germination when treated with smoke.

Bacteria and fungi

Although not plants, these organisms are often overlooked and need mentioning (a longer discussion of the effect of fire on soil can be found in Chapter 6). Bacteria and fungi are remarkably important components of soils in that they decompose dead material and, in the case of fungi, form beneficial relationships with the roots of plants (mycorrhizas). Yet few studies have looked at the effect of fire on these organisms. One study in the pine forests of southern Japan (Mabuhay *et al.* 2003) has shown that the numbers of both bacteria and fungi declined immediately after a fire and did not increase to pre-burn levels for a number of years. This appears to be the norm. A similar pattern is undoubtedly shown by lichens: a dramatic decline after a fire with the likelihood of a long-term increase in species diversity as dead wood created by the fire is colonised.

Animals and fire

Compared with plants, animals have the advantage that they can move away from a fire, but unlike plants most animals are not 'modular', and the injury of one part may lead directly or indirectly to death; a deer cannot just grow a new leg like a tree can grow a new branch. Animals are also more affected by fire changing their habitat and food supply, at least in the short term.

Before looking at how fire affects animals, it is worth pointing out that animals can have a direct effect on fires as well. Mass insect attacks may produce dead and dying trees, and hence fuel. This is clearly seen in eastern Canada with the death of the balsam fir (*Abies balsamea*) caused by the spruce budworm (*Choristoneura fumiferana*) (Taylor & MacLean 2007). Mammals can also have an effect on fire. Beavers influence fuel by felling trees and altering hydrology, and large grazing animals like deer can break up blocks of fuel by intensive localised grazing, and create fire breaks by the creation of trails. Kramer and colleagues (2003), using computer models, predicted that in mixed forests of the Netherlands grazing reduces the occurrence of both small- and large-scale wildfires by reducing fuel load, and altering the vertical structure of the vegetation. Squirrels can also alter the incidence of fire by mutilating the tops of trees, leaving exposed dead wood that is more vulnerable to ignition by lightning strike.

What kinds of animals are killed during a fire?

Insects and other invertebrates face a mixed fate in the face of fire. Mobile insects, such as grasshoppers, spiders and cockroaches, can readily move way from even fairly fast-moving fires. Studies have shown that as many as 80% of individuals can still be alive the day after the fire (but bear in mind that mortality can continue in a 'shock phase' after the fire due to exposure, starvation etc.). Less mobile species or ground-dwellers such as ants are more likely to take shelter in refugia, dropping to the ground, crawling below litter, into rock crevices etc. Intuitively, one would expect the bigger, more mobile individuals to survive best by escaping the fire, but the limited studies so far have shown that it is often the smallest individuals that survive. Larger ones fly strongly but are more likely to be caught by the flames while smaller individuals are more likely to find safe hiding places and survive. A number of invertebrates have evolved ways of detecting approaching fires. It is reported in South Africa that ticks drop to the ground and seek shelter at the slightest detection of smoke, reacting it seems to volatile substances only released at high temperatures.

Insects living on and near the surface of trees will probably be decimated by fire since they are most likely to be exposed to heat, flames and smoke, unless sheltered in nooks and crannies. Snails do not usually do too well either (Kiss & Magnin 2003). However, even the most vulnerable species rarely meet 100% mortality, although it can be close. Spider populations, for example, have been seen to reach 99% mortality in Europe (Heliövarra & Väisänen 1984). Even those populations that are extensively thinned out by a fire can usually recover quickly (providing the habitat has not changed too much – see below). Most invertebrates go through boom and bust cycles as a normal part of their lives; fire adds another bust part of a cycle.

Cold-blooded vertebrates such as reptiles and fish may appear to fare poorly in fire. For example, William Bartram in Florida in the late eighteenth century wrote in his journal about watching king vultures feeding on the roasted serpents, frogs and lizards. Having said this, a large number of studies in North America, tropical forests, European Mediterranean forests and Australian eucalypt forests have found no significant long-term differences in reptile numbers (newts, salamanders, frogs and snakes), certainly after a single fire (Greenberg & Waldrop 2008, Lindenmayer *et al.* 2008b, Masterson *et al.* 2008). These animals generally escaped fires by hiding below ground, under rocks and in moist woody material. Fish are unlikely to be harmed directly by the heat of a fire unless the water body is small or particularly shallow.

But fish can be harmed indirectly by removing shade and allowing water bodies to warm up (and thus have lower oxygen content – trout, for example, rapidly die in such circumstances), and by increased soil erosion and nutrient inputs changing the water quality, and changes in water flow (Burton 2005, Lyon & O'Connor 2008).

Most people are undoubtedly concerned with birds and mammals and whether Bambi ends up gently barbecued. The perceived wisdom is one of nonchalance in the face of fire; deer standing in rivers (Fig. 5.10); caribou (*Rangifer tarandus*) resting on the ground while encircled by fire. Since fires are rarely uniform sheets of fire, larger animals and birds can escape into non-flammable areas (such as wetlands), break through the fire line into areas already burnt or cross fire breaks such as rivers. During the 'disastrous' fires of 1988 in Yellowstone (Box 6.2), only 1% of the elk (*Cervus canadensis*) population died (more died in the following winter but it was attributed to it being a particularly bad one; Baskin 1999). In fact, many animals exploit the fire. In North America, Australia and Africa, insectivorous birds and even raptors are attracted to prey on the insects and small mammals fleeing a fire. In South Africa, hornbills and the cattle egrets are regularly seen patrolling the flanks of a fire to catch fleeing insects. On African savannahs, lion, leopard and cheetah are also seen hunting alongside fires.

Figure 5.10
A classic picture taken on 6 August 2000 of two elk (*Cervus canadensis*) standing in the Bitterroot River, Montana with no apparent signs of panic while an intense forest fire burns behind them. Photograph taken by John McColgan, US Department of Interior, Bureau of Land Management.

However, most observations of animal responses and deaths are from fairly gently controlled fires, since these are the ones with most people present. Inevitably a large fast fire with a continuous front will cause frantic flight and higher mortality. Whelan (1995) repeats a story of locals in Florida lining up along roads to shoot animals running before an advancing wildfire because animals with smouldering fur acted as 'firebrands' setting new fires across the fire-break road. Moreover, the aboriginals of North America and Australia learnt to set fires to drive and cook small game (see Chapter 2).

As with insects, smaller mammals will shelter below ground where soil insulation and their dense fur give them some protection against a direct heat pulse. Experiments using caged animals in the path of fires (back in the days when this was considered acceptable) have shown that death while sheltering underground is usually caused directly by suffocation (due to removal of oxygen by the fire and inhalation of smoke) and by overheating. Heat penetrating into burrows evaporates soil moisture leading to very humid, hot air. Air temperatures above 60 °C at 22% relative humidity or above 45 °C at 82% relative humidity are enough to cause death.

Young mammals and birds are likely to suffer higher mortality if they have limited mobility – like chicks in a nest – or if their response to danger is to lie still – as in deer. But over evolutionary time, the breeding periods of animals in fire-prone areas have been synchronised to avoid the main fire season(s). Some animals, however, are enigmatic. The koala (*Phascolarctos cinereus*) of Australia lives entirely in flammable eucalyptus trees and the world press regularly feature pictures of koalas with burnt paws that have been saved from bushfires such as the one in Fig. 5.11. It is obvious that the koalas do suffer: BBC News reporting on bushfires in New South Wales on Tuesday 23 December 1997 quote Sue Brookhouse, of a local Wildlife Rescue Centre, as saying:

> One koala's pads have been burnt off. The skin on her nose dropped off this morning. The edges of her ears will go. Her eyes have been burnt and she had a baby in her pouch which didn't survive because she stopped lactating.

What is most amazing, however, is that koalas survive at all. They have probably survived in the past mostly in areas where surface fires were the norm, allowing them protection in the canopy tops, and where frequent fires produced a mosaic constraining new fires into a small area. Now koalas are restricted to a mosaic of forest fragments, which helps to prevent fires spreading to all parts of the population (Lunney *et al.* 2004).

The bottom line is that animals are killed by fires. In general, however, it seems that apart from insects, the numbers killed by fires every few years, decades or centuries are normally fairly small compared with the annual loss of numbers through disease, predation and food shortage. What is probably far more important to the long-term health of animal populations is what happens after a fire has finished burning.

Figure 5.11
A juvenile koala that escaped a bushfire in Victoria, Australia. Its ears and nose have been singed by the fire and its paws have been bandaged. Koalas can be hurt even when not directly touched by the flames if they try to climb a tree where the bark is extremely hot from radiative heat from the fire. This one was likely rescued before the fire became too intense, probably on the outskirts of a fire. Koalas that survive a fire can often starve to death because it can be many weeks before new growth appears on the trees. Photograph courtesy of the Australian Koala Foundation, Brisbane (www.savethekoala.com).

What happens to animal numbers after a fire?

Fires cause great changes to the forest or bush in terms of:

- shade, openness and a mosaic of burnt/unburnt areas;
- changes in temperature with a hot black surface, more exposure to the sun with early spring snow melt and rapid spring greening;
- reductions in litter and shelter near the ground but a greater amount of dead material; and
- changes in food supply for herbivores (more palatable green growth but a favoured food plant may be missing) and for carnivores (again, this may be positive or negative depending on food needs).

Thus, animal populations will change to reflect the fire's influence. Some populations usually decline immediately after the fire – soil fauna (including earthworms) and ants usually decline due to drier conditions and loss of litter – while others will

Figure 5.12
A series of differently coloured individuals of the pygmy grasshopper *Tetrix subulata* from south-central Sweden, ranging from black to grey to mottled brown. After a fire in a clear-cut field, the black individuals were the commonest but over the next 3 years a more even distribution of individuals developed across the nine alternative colour types. From Karlsson *et al.* (2008) with kind permission of Springer Science + Business Media.

increase by better reproduction and survival, or by immigration (e.g. Moretti *et al.* 2006). Grasshoppers, moths, butterflies, beetles (especially predatory ground beetles) and many other insects do very well. There are also many examples around the world of birds attracted to burnt areas by increased (or more easily caught) food supplies and subsequent changes in bird populations as the forest recovers from fire (e.g. Green & Sanecki 2006, Schieck & Song 2006, Haney *et al.* 2008, Fontaine *et al.* 2009). Similarly, small and large mammals (from deer and impala to rhinoceroses and elephants) may congregate on burnt areas due to the more nutritious plant growth (e.g. Zwolak & Foresman 2007, Amacher *et al.* 2008, Long *et al.* 2008). But nature can compensate; there are a number of cases of grasshoppers and other insects that show 'fire-melanism', the ability to turn darker when on burnt areas and so be less conspicuous (Fig. 5.12).

Some herbivores (big and small – from deer down to rats and mice) use recently burnt areas as salt licks, eating the ash or burnt plant parts for the high mineral content, and may be seen to eat the bark of burnt trees, which turns out to be higher in protein than unburnt bark. They may also wallow in ashes to rid themselves of ticks and fleas.

A number of insects are attracted to burnt areas for breeding purposes. Several dozen insect species in a variety of families worldwide have been seen to be attracted to fire or smoke, and are so keen that they will land on stumps that are still smouldering (presumably a strategy for getting ahead of competitors). Some, including butterflies and dragonflies, may be using the smoke plume simply as a marker for attracting members of the same species together as a mating swarm without necessarily being interested in the burnt site. Others are attracted to breed and lay eggs in the dead wood. Most of these are insects that bore into dead, dying or fallen wood or other vegetation, or depend on species requiring such substrates (e.g. fungi). These are primarily beetles (particularly in the families Buprestidae, Cerambycidae and Silphidae) but also includes 'smoke-flies' (such as the numerous *Melanophila* species). Many of these insects have infrared detectors (Schmitz *et al.* 2007) that can sense the heat of a fire up to 150 km away, and have wax-secreting glands to protect their bodies against the dry desiccating conditions of burnt wood fully exposed to the sun. Several types of bark beetles (Scolytidae) are attracted to lightning-struck trees by the volatile resins released from ruptured wood and bark. What happens to these insects between fires? They seem capable of hanging on by using trees killed or seriously weakened by wind, lightning, disease and drought.

On any one spot there is normally a sequence of species taking advantage of different conditions as the forest recovers after a fire. As an example, in the forests of northern Canada, after a fire, deer mice (*Peromyscus maniculatus*), moose (*Alces alces*) and black bears (*Ursus americanus*) are typical of early post-fire stages. The white-crowned sparrow (*Zonotrichia leucophrys*) and alder flycatcher (*Empidonax alnorum*) appear after the dead trees have fallen, and northern red-backed voles (*Clethrionomys gapperi*) and caribou (*Rangifer tarandus*) – which eat lichen – are more typical of late fire stages more than 50 years old.

Post-fire recovery of plants and animals

Looking at the bigger picture of all species in a piece of forest, there is usually a pattern in numbers after a fire. Overall numbers of animals will usually dip (sometime quite dramatically) for a few months to up to 3 years after the fire, to be followed by a boom in numbers as the original animal species recover and new ones move in to take advantage of the temporarily changed conditions. This is helped by the mosaic of differently burnt areas – some unburnt and others burnt at different intensities – which enables more species to find suitable conditions in a smaller area (see Chapter 6 for more detail about fire and biodiversity). This post-fire boom is usually then followed by a steady decline to pre-fire numbers (which may take decades) as the forest reverts back to its pre-fire condition. Plants tend to go through the same cycle. Herbaceous species increase in number after a fire, in response to more light, and decline again as the trees reassert themselves.

There can be complications in what appears to be a neat picture where fire benefits all. For example, before a set of large fires in the Canadian Rocky Mountains, elk (*Cervus canadensis*) were uncommon whereas mule deer (*Odocoileus hemionus*), moose (*Alces alces*) and bighorn sheep (*Ovis canadensis*) were common. After the fires, the elk increased in number, preferring the open areas, and the larger numbers of elk tended to move into habitat used by the other three species. Elk compete strongly with all three species for food (and also compete for shelter with the mule deer) and so they all declined in number as the elk increased.

Just how plants and animals change in number after a fire, how fire influences biodiversity, and thus why fire can be used as a management tool to manipulate numbers, is the theme of the next chapter.

6 | The benefits of fire and its use as a landscape tool

With Peter Hobson

Fire and biodiversity – an overview

The region within the European Union classified in the Habitats Directive under category 90 'Forests of Boreal Europe' is but a fraction of an immense biogeographical zone that stretches east, covering over 700 million hectares of Siberia, and that continues on in a vast swathe through northern North America and Canada. Russia alone supports 22% of the world's forests and between 70–75% of this area remains close to natural. The 'Taiga', as it is also known, is large in nearly all senses of the word; the size of habitats, including mires, lakes and rivers that nest within it; the large, roaming herds of big game; the size of the big animals, including moose and brown bear; and not least of all, the scale of disturbances and processes that give this ecosystem its distinctive character. To most who know it, the boreal biome remains one of the last frontiers of wilderness, a 'self-willed' land shaped by the forces of wind, snow, pathogens, herbivores and especially fire.

South of the equator, on the African continent, stretching 2500 km from east to west and 1250 km north to south, lies one of the greatest expanses of uniform vegetation in Africa, the Miombo Forest. This landscape of trees and grass, sometimes dense uniform forest, is dominated by trees belonging to the genera *Brachystegia* and *Julbernadia*. Rather like the boreal forests of the north, the very survival of this vast ecosystem is dependent on cycles of fire that either creep or sweep through it.

Both systems, whilst so different in many ways, also have much in common. They constitute a significant proportion of the world's last remaining areas of wilderness. Large tracts of old growth forest still survive, and they are home to some of our most charismatic species of wildlife. In sheer size and biodiversity both biomes rank alongside the Amazon forest. Not only do these regions have large core areas that protect wildlife from the impacts of human activities, but they are also two of the planet's few intact natural areas still big enough to buffer some of the catastrophic global changes that have occurred in the last 200 years. Old forests store more carbon than any other terrestrial system and the Stern Report on the economics of climate change, produced for the British Government in 2006, called for the protection of natural forests because of the role they play in the carbon cycle.

Fire and biodiversity would appear to conflict: fire kills. Yet without fire, these particular forest systems, as we understand them, would cease to exist. The rejuvenating quality of fire provides new opportunities for colonisation by both plants and animals, and ensures the long-term persistence of biodiversity (Granstrom 2001, Ryan 2002).

Forest fires in the northern hemisphere create shifting mosaics of vegetation of different ages and structure that produce something of a kaleidoscope effect over extended periods of time. On the small scale, stands of pine or spruce can appear uniform since they look much the same wherever you stand in them, whilst a large-scale perspective reveals a complex mosaic of patches of age and structure (see Box 6.1). The rather unpredictable relationship between forests and fire creates the stability that is essential for biodiversity and resilience in the ecosystem. In the absence of fires, species diversity and communities will alter and large-scale spatial heterogeneity will decrease (Granstrom 2001).

Despite stark geographical and climatic differences between two of the largest fire-prone forest biomes on earth, the boreal and miombo systems, the ecology of both systems is dictated by a relatively short growing season interrupted by a long period of quiescence brought on by drought or cold conditions. These climatic patterns have encouraged the evolution of communities that concentrate their activities into a small event window. For the remainder of the year both systems appear to be bereft of life. This is a false impression for in fact life in the forest is in

---

Box 6.1. Biodiversity and disturbance: the Intermediate Disturbance Hypothesis

The Intermediate Disturbance Hypothesis (IDH) states that diversity is highest where there is some disturbance and lowest where there is either very little or too much disturbance. In other words, high diversity is promoted by some disturbance that keeps the dominant species from taking over, allowing a mix of species including those that like open disturbed areas. This has been shown to be true for a number of systems, particularly marine, but a number of high-quality studies have concluded that the IDH does not apply with fire (e.g. Collins et al. 1995, Li et al. 2004). These studies have generally used small plots burnt at different times. A good study by Schwilk et al. (1997) illustrates this. At a scale of 1 m^2 diversity increases with fire frequency, probably because very frequent fires tend to select for small plants and more small plants of different species can be packed into a small area. This high diversity tends to decline with time, so more frequent fires keep the diversity constantly high. Maximum diversity was found with the highest degree of repeated disturbance over the same area, not at intermediate, less frequent amounts of fire.

The argument over IDH, however, really hinges on spatial heterogeneity, the pattern of areas burnt at different intensities and at different frequencies, including areas that may experience a lot of fire and others that may remain unburnt for a long time. This enables more species to find suitable conditions within the landscape. In small plots a fire is normally fairly even across the whole area, that is, there is very little spatial heterogeneity. Even at the largest scale used by Schwilk et al. (0.1 ha or 1000 m^2 – an area 50 × 20 m) the amount of spatial heterogeneity was presumably still fairly small compared with the much greater mosaic on the larger landscape, and so the results are a reflection of how species cope with more or less frequent fires. To put this into more technical terms, following a fire the alpha (α) diversity – that is, the local species richness within a small area – does not appear to follow the IDH, but the beta (β) diversity – the species richness across the landscape – undoubtedly is increased, as more areas are at different stages of post-fire change and so can hold more species with different needs. Biodiversity at the landscape level does seem to fit well with the intermediate disturbance hypothesis. See Whittaker et al. (2001) for more discussion on biodiversity at different scales.

storage for the next wakening event. It is this predictable cycle of activities that gives both biomes their resilience which, in turn, provides ecosystem stability to global regions.

The second dominating feature of boreal and miombo forests is the characteristically poor soils. Miombo forests sit on infertile, free-draining soils whilst conditions in the boreal forests are a complex pattern of peat, waterlogged ground, gravels and permafrost overlying unyielding bedrock. Both systems are largely reliant on the nutrients released from burning which would otherwise remain locked up for long periods of time. In these forests of short seasonal plenty certain species build up in numbers creating annual plagues, whether it is the tsetse fly (*Glossina morsitans*) of the miombo or mosquito (*Heimo culicidae*) of the boreal. Both biomes also characteristically experience a dramatic influx of migrants from outside the region that benefit partly from these biological pulses. Canada's boreal forest provides the summer range for one-third of North America's songbirds and three-quarters of its waterfowl. However, this burst of intense activity is quickly over and the systems appear to retreat into the familiar protracted period of dormancy.

Ecosystems that operate to predictable seasonal or cyclical patterns can, over time, become uniform in how they look and how they function. However, the sporadic and indeterministic occurrence of wildfire breaks up this uniformity, producing patchiness across all scales and new opportunities for species to colonise and persist (see Box 6.1). The landscapes are spatially heterogeneous with complex multi-storied tree canopies. The uneven age structure between stands show reverse-J tree diameter distributions (many small trees and few large ones), and include trees reaching their maximum age. An important component of these systems is the presence of standing dead trees and coarse woody debris in all stages of decomposition. This, together with the multi-cohort character of the live trees contributes to the quasi-equilibrium state of old growth, fire-prone forests where despite local disturbances by fire, the forest at the landscape level is maintained almost unchanged.

Many species in fire-prone forests readily exploit the opportunities presented by wildfire. Birds will take advantage of easy pickings where prey is flushed out by fire or, like the herbivorous mammals, will move in to recent post-fire sites to feed on new growth and the insects this attracts. On any one spot the forest community is normally a relay of species exploiting the diversity of conditions found on a site during the gradual process of recovery following a fire event. As an example, in the forests of northern Canada, after a fire, deer mice (*Peromyscus maniculatus*), moose and black bear are typical of early post-fire stages. The white-crowned sparrow (*Zonotrichia leucophrys*) and alder flycatcher (*Empidonax alnorum*) appear after the dead trees have fallen, and northern red-backed voles (*Clethrionomys gapperi*) and caribou (*Rangifer tarandus* – which eat lichen) are more typical of late stages more than 50 years old (Fortin *et al.* 1995).

Fire also alters and reconfigurates the woody biomass thus presenting many specialist species with the necessary resource to survive. After a fire in the boreal forest, many trees can be left severely damaged or standing dead. This new resource is quickly exploited by the bark beetle *Scolytus ratzeburgi*. Alongside the beetle is the fungus *Fomitopsis pinicola*, a species that causes heart rot in pine and ultimately can result in the tree snapping due to rot at the base (butt rot). The rapid colonisation of a stand by these two species following on from a fire creates something of a biological cascade effect. Ants, either *Formica* or *Myrmica* species, readily take up residence in pine that is suffering from butt rot. The nests of these species attract the attention of the black woodpecker (*Dryocopus martius*) that will drill into the base of the tree to feed on the ant brood chambers. The three-toed woodpecker will systematically work trees infested with *Scolytus*, drilling and bark peeling as it moves through a stand. In some cases the whole tree may be cleaned of its bark. Where the horsehair lichen (*Bryoria capillaris*) remains attached to the dropped bark the reindeer will pick this off (Fig. 6.1). There must be many examples of relationships of this kind that have their origins in forest fire.

The complex relationship between fire and the landscape creates a variation in resources at a large scale which, in turn, has a profound effect on the distribution, function and persistence of biodiversity. In forests that experience periodic intense fires, lichens and mosses are often completely destroyed and replaced by grasses. The release of nutrients in ash and smoke provides new opportunities for many species. The general increased productivity of plants provides a rich food resource for both bird and mammal browsers. Other species to benefit from fire-induced regime shifts include saproxylic groups – fungi and invertebrates that feed on dead wood (or xylem), as well as the dead-wood foragers that feed on these pioneers. Fire sites are also readily colonised by opportunistic predator species including ground beetles (family Carabidae), wolf spiders and money spiders (Lycosidae and Lyniphyiidae).

In contrast, more fire-resistant sites (typically wet pine or spruce stands in boreal forests) that have much longer fire-return intervals of up to 300 years offer many species a sanctuary from the temporarily destructive forces of a fire. Such areas are convenient breeding grounds and safe sites for plants and animals, functioning as biodiversity silos in a landscape of unpredictable change. These fire-resistant ecosystems are vital to the recovery of forest biodiversity as many of the species held in these sites will move back out into the recently burnt stands. Such areas also provide the forests with the necessary resilience to withstand the long-term effects of periodic burning.

Beyond our broad understanding of the role of fire in promoting forest biodiversity it is very difficult to establish quantifiable evidence of how this complex relationship works in detail. Fire regimes have altered considerably over time in response to the effects of a changing climate, and, more recently, through the dramatic influence of humans. Consequently, it is unlikely that any patterns of

Figure 6.1
Bark debris left scattered around the base of a dead Scots pine tree (*Pinus sylvestris*) as evidence of foraging activities of the three-toed woodpecker (*Picoides trydactylus*). Bark shavings that are still covered in lichen are often grazed by reindeer during the winter months. Photograph by P. Hobson.

existing biodiversity will correlate accurately to present-day fire regimes since they are a result of previous and often unknown fire regimes. Without knowledge of pre-human disturbance any attempt to quantifiably link biodiversity and fire regimes are at best fuzzy. The picture may be even more complicated in certain taxonomic groups (in particular, the invertebrates) because of their normal extreme variation in numbers over small distances and over short periods of time. This variability is attributed more to changes in environmental factors such as temperature, light and moisture which may mask influences of fire disturbance. For instance, some groups of invertebrates may increase in abundance after a fire while others decline, and yet, the responses of these particular groups may not be the same after the next fire. For a number of insect groups, including the flies and beetles, large seasonal fluctuations appear to outweigh any fire-related trends.

Finally, it cannot be assumed that recolonisation of post-burn sites by invertebrates will recover to the original pre-fire state.

Looking at the landscape scale, certain patterns in fire-prone forests are revealed that may provide further evidence of the complex relationship between fire and biodiversity. The association of fire disturbance with the quantity of dead wood in northern coniferous forests is well documented, and it is the generation, quantity and dynamics of this particular resource that affect so many species populations. This is discussed below.

Unpicking the factors that affect biodiversity

As described above a large number of factors affect the ecological aftermath of forest fires. Although these often interact in a very complex manner, it is worth singling out the main features of the fire regime of an area (see Chapter 3) and looking at how each of these affects biodiversity in more detail.

Fire frequency

How often a fire reburns an area (for the ways in which fire frequency or interval can be defined, see Box 4.7) is critically important to the sort of forest that grows on a site. If fire reburns an area before plants have developed fire-proof properties (such as thick bark – see Chapter 5) or before they have started producing and/or storing seeds, these species will eventually be lost. Conversely, if fire becomes rarer, those species that require fire to persist or reproduce will be lost. The important factor is the length of time between fires and the time that a species can persist without fire.

These patterns of species-change with different fire frequencies have been extensively studied. Díaz-Delgado and colleagues (2002) looked at Mediterranean shrubs in north-east Spain using satellite pictures and found that in areas burnt twice, vegetation recovery within 6 years was lower after the second fire than after the first. Recovery after the second fire gradually improved the longer the time since the first fire; this suggests that although the shrubs are fire-adapted the whole ecosystem takes time to recover and is affected if fires come too frequently. In other words, increased fire frequency reduces ecosystem resilience, that is, the ability to recover to its pre-fire state. Forests dominated by oaks that sprouted from below ground were more resilient to the first fire, but showed a greater decrease in resilience after the second fire than did forests dominated by pines that regenerated from seed.

The same team carried on this work by using computer models to work out what would happen in this area if fire frequency changed dramatically; some of their results are shown in Fig. 6.2. They used a computer model to predict how vegetation would change over a 200-year period with different fire frequencies.

Figure 6.2 (opposite page) How do different fire frequencies affect the shrubby Mediterranean vegetation near Barcelona in Spain? This figure results from a computer simulation over a 200-year period for two species – the shrubby kermes oak (*Quercus coccifera* – shown below at Kusadasi, Turkey) and a heather (*Erica multiflora*) – under different fire frequencies from a control of no fire to fires every 100 to every 5 years. The data show how long each of five life stages lasts under these different frequencies. So, for example, with no fire, the kermes oak would exist as abundant adult plants for 43 years and as scattered mature plants for 20 years whereas with fire every 5 years, mature adult plants would be found for only 20–30 years over a 200-year period. Figure from Lloret *et al.* (2003) Fig. 2. with kind permission from Springer Science + Business Media.

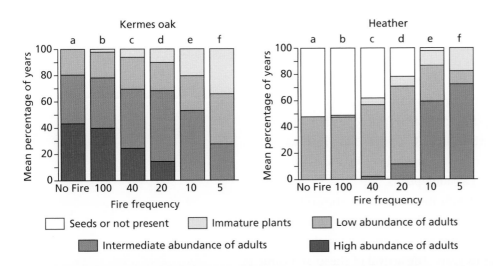

Six frequencies were used from no fire to a fire every 5 years on average (seen along the bottom of the graphs). The abundance of each plant was estimated as the average per cent of years out of 200 years that each of the five stages of the plant were present, from seed to adult plants. As fire becomes more frequent, the shrubby kermes oak becomes less dominant, the mature trees are progressively killed above ground and recover by sprouting and so the adults are replaced by immature plants that do not have time to mature between fires. By contrast, the heather, which grows to maturity more quickly, and is normally shaded out by the oaks, prospers under frequent fire with more time spent as adult plants.

Fire frequency does, of course, interact with fire intensity as shown by the fate of Ponderosa pine stands (*Pinus ponderosa*) in the American south-west. Early explorers found open stands with grasses that reached the bellies of their horses and through which fire burnt every 2–15 years. The arrival of sheep and cattle in the 1800s soon saw the disappearance of that grass. On the whole, this was seen as a good thing since it cut fuels and reduced the incidence of fire. But as the grass disappeared there were tremendous problems of erosion with gullies up to 15 m deep, and the pines increased from fewer than 50 trees/ha to more than 2000/ha. Animals such as grasshoppers, bluebirds and pronghorn antelope that needed the grasses declined and others that needed dense wooded areas, such as porcupines, increased (McInnis 1997). Fires became much less frequent but when they did start they were intense, often crowning and moving fast through the accumulated fuel, killing many of the old trees that would have survived the light grassland fires.

Further north in Glacier National Park, Montana, ponderosa pine has met a similar fate. Here the fire frequency was around 13–58 years before European settlement, but it is now around 69 years (Ryan & Frandsen 1991). This has led to an accumulation of litter (needles and bark) around the base of the pine trunks which leads to more severe fires there and greater mortality. Bear in mind, however, that other forests have long had low-frequency but high-intensity and severity fires, and thrive on them; Box 6.2 describes just such a forest in Yellowstone National Park.

At its most extreme, fire frequencies that are too high can prevent forest regrowing and it is instead replaced by 'pyrophytic' grassland. This has occurred in parts of Indonesia, Amazonia and in North Queensland in Australia due to tree-felling/agricultural clearance which allows drying of the remaining vegetation, all of which is aided by droughts from El Niño. In these cases the dense evergreen forest degenerates to a few fire-resistant trees surrounded by a low cover of weedy species (Dennis *et al.* 2001). Such changes can, however, happen in any fire-prone forest as well if the fire frequency is high enough.

In many forest systems wildfire frequency can vary considerably between stands depending on the local conditions. For instance, in northern European boreal forests pine growth on free-draining sandy soils and south-facing slopes is likely

Box 6.2. Fire in Yellowstone National Park

Yellowstone National Park in north-west Wyoming, USA was the world's first National Park, founded in 1872. A census in 1890 reported dwindling timber supplies and this led to the Forest Reserves Act of 1891. Foresters such as Gifford Pinchot pushed for scientific management of designated National Forests as a means of conserving timber. At the same time John Muir popularised the recreational role of wild places as a relief from urban and industrial civilisation. Although at odds with each other, both reflected human rather than environmental priorities and helped reinforce the idea that fire suppression in National Parks was a good thing.

The lack of fire, however, gradually led to the forests becoming dense, drab and with fewer species (photo, below right). Even the elk (*Cervus canadensis*), a feature of visits to the park, began to go to the more open lands elsewhere and became more elusive. Ecologists began to doubt the wisdom of fire-suppression policies, recognising the natural role of fire in nutrient circulation, removing dead wood and maintaining biodiversity. Starting in 1972, American National Parks initiated a natural-burn fire policy. The new 'let-burn' policy allowed naturally occurring fires (i.e. lightning-started) to run their course over about 15% of the park's natural area provided they did not pose a threat to visitor areas. Between 1972 and 1987, only 14 000 ha of Yellowstone's 900 000 ha (1.6%) of grasslands and forests were burnt in 235 fires. Just 15 of those were larger than 40 ha (100 acres) and all went out without help. The summers of 1982–1987 were wetter than usual which undoubtedly helped.

1988 was a little different! April to May was wetter than normal but from June on, the whole area experienced drought and the summer was the driest in the park's history. Dry lightning storms were common and by 21 July a decision was made to suppress all fires but the fires carried on burning. By the end of the next week, 40 000 ha had burnt and the fires continued until 11 September

1988, when the first snows came. By this time, there had been a total of 50 fires that impinged on Yellowstone National Park. Of these seven major fires burned 95% of the burnt area within the park; five of these started outside the park and three were human-caused. More than 25 000 firefighters battled the fires at a cost of US$120 million. By the end, 312 000 ha (c. 36%) of the park was burnt.

The media had a field day and told of the jewel in the National Park's crown destroyed by fire. But these fires did not destroy the park, they rejuvenated it. Surveys suggested that < 1% of soils were heated enough to burn below-ground plant seeds and roots, and immediately after the fire a diverse vegetation grew and lodgepole pine (*Pinus contorta*) regenerated vigorously. Surveys of dead animals after the fires also found just 345 elk (of an estimated 40 000–50 000), 36 deer, 12 moose (*Alces alces*), six black bears (*Ursus americanus*) and nine bison (*Bison bison*) in the greater Yellowstone area of 500 000 ha. Most of these were killed as fire swept quickly down two drainages. Some small fish were killed due to either heated water or fire retardant dropping in streams, but the animal populations recovered quickly. Certainly, 20+ years later, Yellowstone has recovered rapidly and is in better shape than it has been for a long time (Schoennagel *et al*. 2008).

Due to its geographical position and climate, Yellowstone National Park tends to burn intensely and extensively or not at all. This is helped by the lodgepole pine burning poorly until it is 250–300 years old; before that there is a lot of green material in the understorey and the pine canopies are spread well apart and fires spread slowly. Fires are small and infrequent until the right conditions occur and then large fires scour the landscape. Fire scars show that major fires covered 19% of the park between 1690 and 1709, and 15% between 1730 and 1749.

More details can be found in McInnis (1997), Baskin (1999) and on the National Park Service website (www.nps.gov).

to have shorter fire intervals of 80–150 years compared with similar stand types on level ground or hilltops with heavier soils. These stands may have extended periods of 150–300 years between fires. As a result a forest can take on a distinctive patchy character. The drier, more frequently disturbed areas of forest are dominated by vegetation that is more tolerant of fire, and includes almost monoculture stands of pine, an abundance of heather (*Calluna vulgaris*), cowberry (*Vaccinium vitis-idaea*) and lichen species – *Cladina* and *Cladonia*. More moist and less disturbed pine forest supports a greater diversity of trees in the canopy that includes birch (*Betula*) and spruce (*Picea*). This diversity is also reflected in a more robust field layer with greater amounts of ericaceous plants such as bilberry (*Vaccinium myrtillus*), crowberry (*Empetrum nigrum*) and other woody shrubs including Labradorean tea (*Ledum* species) and junipers (*Juniperus* species). The ground cover of lichen is replaced by a deeper bed of moss, dominated by *Pleurozium schreberii*.

Intensity and severity

As described in Chapter 5, both fire intensity (the energy output above ground) and severity (how far it burns beneath the litter into the peaty soil and dead wood) affect the survival of plants and how much they regrow after fire. It is usually the case that higher fire intensity and severity lead to lower plant cover immediately after the fire due to greater mortality of plants and seeds below ground. However, subsequent performance of the vegetation may be better. For example, studies in Spain by Pausas *et al.* (2003) showed that aleppo pine seedlings (*Pinus halepensis*) were just as dense on areas burnt at different intensities, but survival and growth were better on more intensely and severely burnt sites. This is because, they suggest, on intensely burnt areas, more of the organic matter was turned to ash, readily taken up as fertiliser, and the high severity reduced competition from sprouting grasses and herbs.

Conversely, burning at too low an intensity can impede vegetation recovery. This is due to such factors as inadequate nutrient release in ash and failure in triggering the germination of buried seeds, but the causes can be more subtle. For example, in jarrah forests (*Eucalyptus marginata*) of south-west Australia, nitrogen is very scarce in the soil. Much of the supply of nitrogen comes from the plant called prickly Moses (*Acacia pulchella*) which fixes nitrogen from the atmosphere into biologically usable forms, as do other members of the pea family (Fabaceae). This plant is dependent upon fire for the germination of its seeds, so low-intensity fires lead to fewer prickly Moses plants and so to loss of productivity in other species. Moreover, a high density of the plant makes the site unfavourable for the pathogenic root rot fungus *Phytophthora cinnamomi*, a major cause of dieback of both canopy and understorey species.

Fire intensity in northern boreal forests of Europe plays a key role in determining the succession sequence of plants. The successful regeneration of both pine and birch is dependent on the extent of removal of moss cover and organic matter from the soil surface. Light-touch surface fires may scorch the upper half of the moss layer, leaving much of the rest intact and thus preventing the establishment of these two tree species. On the other hand, spruce and rowan (*Sorbus aucuparia*) are better equipped to establish growth under these conditions, leading to good cover of moss (particularly *Pleurozium schreberii* and *Hylecommium splendens*).

Season

Whether a fire burns in spring or autumn, or any other part of the year, will affect what happens afterwards through such factors as soil and plant moisture, seed availability and, for animals, nesting of birds and timing of vulnerable young. As an example, longleaf pine (*Pinus palustris*) mortality is greater after a spring fire than late summer fires.

Patchiness and animals – does size matter?

Variations in topography, fuel and weather (see Chapter 4) usually ensure heterogeneity or patchiness of a burnt area, from intensely burnt areas creating open habitat through to unburnt islands that shelter fire-intolerant species. Frequent, low-intensity fires tend to lead to a greater within-site heterogeneity whereas large fires tend to homogenise landscapes. This is because trickling fires are much more prone to small variations in a forest while larger fires tend to sweep over these. Nonetheless, larger fires that burn over many days can create a large-scale heterogeneity since the fire will vary in intensity due to daily changes in temperature and humidity and longer-term changes in weather. At the other end of the scale, fire suppression may lead to less patchiness as areas burnt at different times reach maturity and become similar; and when fires do start they are likely to be large and intense. This shows one of the values of prescribed fire, in that small well-controlled fires can help increase patchiness, although these are likely to be all low-intensity fires and so do not perfectly mimic the range of naturally occurring fires.

Patchiness is very important in maintaining high biodiversity since different plants and animals are more likely to find suitable conditions in a patchy forest. Even something as apparently robust as a large forest tree can be affected. In the boreal forests of north-east China, spruce grows only in unburnt refugia even though the burnt areas seem otherwise suitable (Goldammer 1993).

And, yes, size does matter, but it is the size of gaps rather than the overall size of the fire that is generally important. In fact in Australia, Bradstock (2008) points out that large fires are not necessarily any more detrimental than small fires. Large gaps created by fire tend to favour plants that like open, light conditions whereas smaller patches with surrounding trees favour shade-tolerant plants. For instance, birch, an important boreal forest species, struggles to establish a hold in small canopy gaps in spruce-dominated forests. Also, the larger the gap, the less likely that seeds from trees

(a)

(b)

surviving at the edges will reach the centre, but most fires are patchy enough that this is not a common problem (see Fig. 6.3 for an example from Korea). Animals are also sensitive to patch size. The centres of large areas may be unused by timid animals whereas small patches and edges are used since they can readily retreat into unburnt areas. The amount of edge rather than patch size itself may be even more crucial to animals. Even large deer prefer the edge to the centre of a large opening or the unburnt forest. Similarly, species of ant, including wood ants, establish colonies along patch edges, rarely moving into the centres of large clearings.

The influence of patch size on animals often has important effects on the plants. A few small fires may increase herbivore pressure by providing oases of grazing, which in the European boreal forest may be intense enough to eliminate birch seedlings, paving the way for Scots pine (*Pinus sylvestris*) to establish. In larger patches, less herbivore pressure cannot prevent birch establishing which in turn does prevent Scots pine establishing but allows in Norway spruce (*Picea abies*). Grazing may also change what sort of fire burns next. In northern Australia areas heavily grazed during the wet season have less fuel and so burn less intensely during the dry season, leading to shifts in species composition (Andrew 1986).

Figure 6.3
(a) Forests of Japanese red pine (*Pinus densiflora*) 2 years after fires swept through 23 794 ha of forest around Samcheok in eastern Korea. Before the fires along the east coast of Korea in April 2000, 70% of the forest was dominated by Japanese red pine while stands of various oak (*Quercus mongolica, Q. variabilis* and *Q. dentata*) occupied just 3%. After the fires, however, this pattern was reversed (Choung *et al.* 2004). It might initially be thought from (a) that the widespread burn killed so many pines that there were few seeds and so little pine regeneration. In reality, after the fire the oaks and other hardwood trees resprouted so vigorously that the pine seedlings could not compete with the robust oaks. (b) A healthy specimen of daimyo oak (*Quercus dentata*) resprouting with dead pine trees in the background.

Other factors

Predicting the effects of an individual fire or set of fires on biodiversity can be difficult due to the large number of variables that come into play. These include variations in the number of seeds produced each year, rainfall and temperature patterns and herbivory. Fortunately, over many fires these tend to even out so it can be easier to predict on the larger scale. Moreover, Malcolm Gill in Australia suggests that it is the largest fires that matter and these will be at the higher end of the intensity scale and will therefore be more predictable than smaller ones (Binkley 1993).

In addition, as noted at the beginning of this chapter, all these different factors interact with each other. So, in longleaf pine (*Pinus palustris*) stands in south-east USA, oaks can establish and survive mild fires; if they become large they can survive more severe fires and will eventually dominate; if fires are frequent and/or intense enough the oaks cannot establish and longleaf pine dominates. Similarly in Australia, if the mountain ash (*Eucalyptus regnans*) is eliminated by two severe fires in a row, the grassland that develops may resist the invasion of eucalyptus for many years.

Environmental legacies: dead wood and biodiversity

Forest fires affect not just how an area recovers but also the input rate of dead wood (Siitonen 2001). Dead wood (often referred to as coarse woody debris, CWD) is intimately linked to fires, and a single burn has the potential to create a volume equivalent to 100 years' input generated by other causes (Sippola *et al.* 2001). Volumes of coarse woody debris are highest just after fires that kill the trees, and are significantly greater in unmanaged forests where high-intensity fires are more frequent. Dead wood can account for over 50% of the total woody biomass in recently disturbed natural forests in contrast with just 20–30% in equivalent areas where fire disturbance has been absent (Stokland 2001, Thomas & Packham 2007). Furthermore, dead wood lingers in boreal forests; the average retention period of standing dead trees (snags) is 81 years and for logs on the ground it is 118 years.

What is the ecological significance of all this dead wood? In recent years it has been realised just how vital coarse woody debris is in a forest – an estimated 20% of all forest biodiversity is associated in some way with dead wood (Hunter 1990, Grove 2001, Siitonen 2001, Marcot *et al.* 2002). Its value is underlined by looking at beetles that feed on dead wood (saproxylic beetles); they respond to changes in coarse woody debris but not to the relative densities of canopy tree species (Jacobs *et al.* 2007). A study in Finland by Siitonen (2001) estimated that 25% of all forest-dwelling species are dependent on dead-wood habitats and that if natural quantities were reduced by 98% then this would lead to a likely loss of over 50% of the original saproxylic species. Large standing and fallen dead trees provide essential habitats for wood feeders, shelter for vertebrates, protected runways for small mammals, foraging sites, long-term sources of energy and nutrients, and moist refugia during drought or fires (Harmon *et al.* 1986, Berg *et al.* 1994, McCay 2000). Of all the bird species in the boreal forest, 62% are cavity nesters that rely

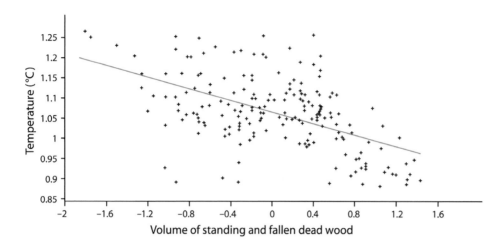

Figure 6.4
The relationship between the volume of standing and fallen dead wood, and fluctuations in local ambient temperature readings (represented here as standard deviation values) in old growth forest, Northern Finland. The trend suggests that as the volume of dead wood increases then there is a corresponding decrease in fluctuations of local ambient temperature.

on the plentiful presence of dead standing wood to provide the right conditions for breeding (Harmon *et al.* 1986).

Dead wood also plays an important part in maintaining the forest's health. On recently disturbed sites, coarse woody debris helps reduce erosion and helps to maintain soil processes (Zielonka & Niklasson 2001). Furthermore, extremes of temperature at the small, microclimatic scale appear to be moderated by the presence and quantity of dead wood (Fig. 6.4). At the very large scale, the accumulation of woody biomass (live and dead) serves as a carbon store to mitigate against climate change (Williams & Gove 2003).

Forestry management tends to decrease the amount of dead wood, and this is the most obvious change when an area of natural forest is first managed for timber production. This reduction poses a threat to the persistence of many saproxylic organisms and the other species that rely on them. The volume of dead wood in Swedish managed forests is estimated at 6.0 m^3/ha in contrast to Fenoscandian old growth, which is between 20 to 130 m^3/ha. Responding to this evidence, the Swedish government has set targets for dead wood in commercial forests to increase by 40% from 1995 to 2010. The evidence for the benefits of dead wood is so compelling that global and national policies for the sustainable management of forests now incorporate a strategy for the promotion and retention of dead wood. This is borne out by European policy for sustainable forest management and is evident in the Improved Pan-European Indicators for Sustainable Forest Management (MCPFE 2002).

Fire, forests and conservation

Wildfire has shaped the structure and boundaries of vegetation in many forests, giving rise to self-regulating ecosystems. Native species and the way they interact (ecological processes) have evolved to depend on conditions created by fire. However, more recently, human-induced changes in fire regimes and forest structure have caused this stable state to shift.

Carl Linnaeus in the eighteenth century described much of Sweden's forests as a wasteful neglect of unproductive forest. At that time approximately 50% of the forested land would have had some evidence of fire, and a substantial proportion would have been old growth. By the mid nineteenth century fire suppression in the Fenoscandian region was proving very effective and has continued through to almost present day, despite recent policy to increase the amount of prescribed burning. Currently, just 0.017% of Sweden's forest burns each year. In addition to active fire suppression, the forests were also systematically converted from a system of low-intensity log extraction to highly organised production. These changes have had a profound effect on the ecology of boreal forests in this region. Where originally 83% of the forest could be described as mostly natural with a diverse age structure, a little less than 0.6% is currently so. Concurrently, there has been a substantial increase in single species stands with a particular loss of broadleaved trees – aspen (*Populus tremula*), willow (*Salix* species) and birch (*Betula*) – which were considered less valuable than the conifers. Furthermore, snags (standing dead trees) and dead wood with logs were removed and a conversion of forests to young and intermediate-aged trees encouraged. The full ecological implications of this change have not yet been fully seen. However, significant losses in fire-associated invertebrate species have been recorded for the region including a number that have gone extinct. The scientific evidence for the decline of biodiversity in boreal forests together with recent studies on the positive benefits to wildlife of fire have convinced policy makers of the need to promote prescribed burning in forestry practice (Fig. 6.5). Since the introduction of the Forest Stewardship Certificate (an international programme to foster sustainable forestry), many countries within the boreal region have introduced national criteria to promote prescribed burning on clear-cut sites in an attempt to make forestry more natural (see also the section below on whether clear-cutting can replace fire).

In Sweden the recommendation is for 5% of annual clear-cut to be burnt although the practical complications of administering this policy substantially reduce this target. A technique of burning slash (logging debris) is also being trialled in a number of forest reserves across Fenoscandia. Investigations into the effects of prescribed burning within the forest indicate that changes to plant diversity strongly mimic natural fire disturbance (Ruokolainen & Salo 2006).

Whilst this evidence is encouraging, there is scepticism about the true likeness of prescribed burning to the original, primeval nature of forest fires. But unfortunately, we do not really know how prescribed fire affected wildfire in natural forest at a large scale in the Fenoscandia region and this forest no longer exists untouched. At best, legacies of historical fires in old human-disturbed forests offer the only near-natural reference sites as benchmarks for good practice. Other concerns about the value of prescribed burning are more to do with the actual practice of burning. For instance, during the 1950s and 1960s managed burns in Fenoscandia were carried out to reduce the amount of fuel in

Figure 6.5
Prescribed burning carried out using a hand torch
in a designated stand of mixed pine and spruce in
northern Finland. This particular fire was one of
40 trial sites supervised by the National Parks
authorities in 2005. Photograph by P. Hobson.

commercial forests and thus minimise the incidence of catastrophic fires. The consequence of these more frequent fires was that for a number of native species the fire regime outstripped their ability to recover, so leading to local extinction. Furthermore, sites targeted for prescribed burning rarely coincided with areas of forest that historically burnt frequently, such as dry pine sites. Instead older forest areas with thick layers of humus were typically chosen even though these would more normally be burnt very infrequently, and are more typically regarded as fire refugia. But, of course, these are the safest areas to burn since the fire is likely to be of low intensity and so easily controlled. This practice has led to a number of initiatives to introduce more appropriate fire management into forest reserves.

In Canada the situation is rather different. Approximately 2 million hectares of forest are burnt each year and there is growing concern that this figure is likely to rise as a result of climate change. The response of the forest industry has been to explore other possible alternatives to fire management. Canadian Forest Stewardship Certificate guidelines encourage the use of clear-cutting but leaving some trees standing to emulate natural fire. Moreover, keeping snags and logs at different stages of decay throughout the harvested forest is encouraged. Similarly, clumps of young trees, in particular broadleaved species, are left as wildlife cover and clusters of large canopy trees are retained for nesting birds of prey. Harvesting activities have also been modified to protect biodiversity in commercially managed forests. Typical patterns of homogeneity found in worked forests are broken up by felling tree-groups. The small openings generated this way mimic gap dynamics created naturally by fire, wind or pest infestation.

Can clear-cutting replace fire?

As described above, there is a growing feeling that for forests to retain as much of their biodiversity as possible, forest management should increasingly emulate natural disturbances – the so-called natural disturbance hypothesis. Thus, some trees can be left standing to visually resemble the structure of an area burnt by wildfire. A corollary of this worthy goal is that it may be possible to produce natural-looking, species-rich forests without having to use fire as a management tool. Although fire is a cheap tool to use, it can of course be difficult to control, with dire consequences should a fire escape and damage surrounding property or, even worse, claim lives. Using forestry techniques to produce 'natural' forests will not prevent wildfires but it might help reduce them by breaking up the landscape into a mosaic of small areas, and it would help reduce the number of management fires that escape and run amok.

The arguments for emulating natural disturbance by cutting down trees hinge on whether clear-cutting an area really simulates the effects of fire. The controversy, dating back many decades (e.g. Chapman 1952), has been fuelled by the mistrust by conservationists that this is just a back-door route for foresters to fell more timber, and the feeling by foresters that setting fire to, say, a National Park, is highly dangerous and wastes a huge amount of timber that could be used with no negative effects on the park (Malakoff 2002). Moreover, the proposals are criticised by scientists for being too simplistic as they are based on little more than size of fires, retention of a few snags and a little burning (Schnider 2001); some of these concerns are addressed in Box 6.3.

Box 6.3. Some of the differences that result from burning an area of forest compared with clear-cutting to try to emulate a fire. Adapted from Hammond (1994)

	Fire	Clear-cut
Size and shape	Messy, irregular	Planned, geometrical
Frequency	Irregular, unpredictable	Short, regular cycles
Patchiness	Variable retention of untouched areas	Total removal, retention – crude/poor representation
Soil structure	Soil cracking and generation	Soil compaction and erosion
Nutrients	Increase in P, N, Ca[1]	Nutrient depletion
Habitat	Retains coarse woody debris and snags	Most trees and features removed
Succession	Browse for animals, creation of canopy gaps	Forestry practice minimises succession stages

[1]Phosphorus, nitrogen, calcium.

So where does the truth lie? A large number of studies have looked at the differences between harvested and wildfire sites in places like Australia but especially in the boreal forest of Canada. In many of these studies the differences in wildlife between the two are down to how much woody material is left standing – the more standing trees left, the more the wildlife resembles what was there before the disturbance. But there are still differences. After fire the standing trees tend to be dead while after harvesting the remaining trees tend to be living and in small clumps. Thus in Canada, burnt areas attract more woodpeckers using the dead trees while harvested areas have more birds of open scrub (Hobson & Schieck 1999). Harvested sites also tend to have higher bird diversities immediately after disturbance (Schieck & Song 2006, Stuart-Smith *et al.* 2006). Differences can also be found in mammals: again in Canada, Fisher & Wilkinson (2005) found that juvenile martens (*Martes americana*) used recently burnt areas but adult martens and fishers (*M. pennanti*) avoided young clear-cuts, and moose (*Alces alces*) and deer found more to eat in the rapid shrubby growth of cut sites. Other studies have shown similar patterns for insects such as spiders and beetles, except the burnt sites develop a higher diversity of species quicker than harvested sites (Buddle *et al.* 2000, 2006). Differences also exist in the plants; in Canada, Haeussler & Bergeron (2004) found that burnt sites in Canada had more aspen suckers and, due to the high light levels, more mosses, while harvested areas had a much denser layer of shrubs and herbs. Penman *et al.* (2008) found a similar relationship in the dry eucalypt forests of New South Wales. Perhaps not surprisingly some of the most striking differences are in the soils. In conifer stands of North America it has been seen that fires kill many more bacteria than harvesting does (74% reduction in microbial biomass compared with 18%), and fires produce a large pulse of nutrients (as described above) leaving burnt soils with a lower nitrogen content (LeDuc & Rothstein 2007, Smith *et al.* 2008, Thiffault *et al.* 2008). These differences between harvested and burnt sites have been used by contenders on both sides of the argument to support their goals, and the controversy is not yet settled. It is important to bear in mind, however, that many of the differences between the burnt and harvested areas disappear after a few decades; for soil nitrogen it may be less than a decade and for plants and wildlife between 2–4 decades.

A variation on this problem is now rearing its head: should logging after a fire be allowed, on the assumption that since the timber is dead anyway it might just as well be cleared out of the way and put to good use? Discussion of this is beyond the scope of this book; the reader is directed to articles such as Noss *et al.* (2006) and the excellent book by Lindenmayer and colleagues (2008a), *Salvage Logging and its Ecological Consequences.*

The future for fire-prone forests: environmental uncertainty, macroecology and ecosystem resilience

The last 200 years has marked a profound change in the history of forested landscapes. Most of the world's forests are now being utilised by human societies and management practices have reshaped these ecosystems at all scales, reducing great swathes of once unbroken green canopy to a complex patchwork of wooded islands. At a smaller scale, exploitation of forests has altered forests enough to change the quality of habitat for numerous species. Where we see this change in Europe, 20–50% of mammals and 15–40% of birds have ended up being classed as threatened according to IUCN's red-data book.

Our understanding of how disturbance by humans departs so dramatically from even the most severe forms of natural catastrophe is superficial, and hampered by the lack of solid scientific evidence. However, we do have convincing data for the continuing decline of biodiversity, enough to prompt international conservation to set ambitious targets to attempt to reduce its loss by 2010. So far there is no evidence to suggest that this target is being met in any of the major world biomes.

For many, the conservation of biodiversity is seen as a 'set-aside' issue that requires specific rules, principles and dedicated land to effectively meet its objectives and targets. This principle focuses on the composition of the biodiversity while underplaying the role that different species play within the forest; more is better regardless of whether species that are critical to keep the forest going are lost. Furthermore, it promotes the compartmentalisation of nature, dividing the landscape into conservation reserves and those areas that are of no interest. This marginalises ecological interests outside designated areas, a convenient position for many stakeholders and a concern to conservation organisations that recognise the limitations of trying to meet both global and national targets for reducing biodiversity loss under the existing constraints of the protected areas framework.

The current strategy for conservation is founded on the philosophical principles and environmental ethics developed at the turn of the twentieth century, coupled with more recent scientific progress made in the field of systematic conservation planning (Margules & Pressey 2000). The product of this evolution is a highly efficient strategy that identifies, evaluates and prescribes for biodiversity according to set rules and principles. Ultimately, it leads to the packaging of nature and to its devaluation and perceived importance in the wider landscape matrix. The incentive that drives the conservation sector towards this practice and reluctantly away from a more informed understanding of open, dynamic ecosystems is that it is easier to produce and meet objectives for targets that are spatially restricted to carefully delineated patches of the landscape. Free-ranging biodiversity with all the problems of uncertainty and the indeterministic tendencies it attracts presents all sorts of complicated issues to the management or governance of biodiversity.

Recent shifts in theoretical ecology towards dynamic systems and resilience – the non-equilibrium paradigm – have encouraged the conservation sector to think more about ecosystem processes, the very glue that binds biodiversity together. However, current linear models of scientific investigation and experimentation have struggled to provide enough unequivocal evidence to satisfy the demands of evidence-based conservation. In cases where scientific findings are either inconclusive or fuzzy, increasingly, the precautionary principle is being applied to the decision-making process in the hope that elements of uncertainty are factored into strategies to preserve biodiversity. Researchers are now turning more towards multi-hypotheses modelling. For their part, conservationists are reciprocating by employing adaptive management techniques, and engaging with more innovative practices including mimicking natural processes. This approach is particularly pertinent to systems that are governed by aggressive disturbance regimes of the type seen in fire-prone landscapes. Many of the threatened and specialist species targeted by conservationists are dependent either directly or indirectly on forest fires that operate at a large scale. Disturbances of this magnitude are indiscriminate and generate unpredictable scenarios that, in turn, give rise to complex structures. How are modern-day ecological principles and conservation practices applied to fire-prone forest landscapes?

Working with the new science paradigm, conservation organisations recognise that dynamic processes in nature ignore artificially contrived boundaries on the ground just as species are likely to filter in and out of reserves. This understanding has prompted a rethink in conservation strategies for large and dynamic forest landscapes.

A range of large-scale initiatives including landscape planning of the kind seen in the Natura 2000 programme within the European Union and the wider pan-European initiative of PANParks have provided managers with benchmarks for good practice (Fig. 6.6). At another level, the introduction of forest stewardship initiatives and policies on sustainable forestry outside protected areas has promoted environmental cross-compliance throughout many of the forested regions of the world.

Landscape planning takes a comprehensive, multi-scale, multiple-species approach to management, a strategy that is widely practiced by countries that encompass fire-prone forests. This policy extends to international, trans-boundary efforts between countries that share large tracts of forest. The guiding principles to meet this primary objective include the maintenance of:

Mikonsuonpalon ennallistamispoltto

Mikonsuonpalo sijaitsee Oulangan Natura 2000 -luonnonsuojelualueella (29 390 ha), joka on merkittävä vanhojen metsien ja soiden kokonaisuus. Tulevaisuudessa Mikonsuonpalon alue tullaan liittämään osaksi Oulangan kansallispuistoa.

Metsähallituksen luontopalvelut poltti vuonna 2006 Mikonsuonpalossa seitsemän hehtaaria entistä talousmetsää. Polton valmistelut aloitettiin kevättalvella kaatamalla osa puista hangelle kuivumaan. Polttoalue ympäröitiin puuttomalla palokäytävällä ja läheiselle suolle kaivettiin kaivinkoneella sammutusvesikuoppia. Ennallistamispoltto toteutettiin heinäkuun lopussa.

Poltto onnistui hyvin ja palojäljestä tuli mosaiikkimainen. Alueelle syntyi runsaasti eriasteisesti palanutta, hiiltynyttä ja vähitellen lahoavaa puuainesta, jota monet uhanalaiset palolajit tarvitsevat menestyäkseen. Metsän monimuotoisuutta lisäävän lehtipuuston kehittyminen mahdollistuu humuskerroksen palaessa.

Seurantatutkimukset

Metsäntutkimuslaitoksen Muhoksen toimintayksikkö tutkii Mikonsuonpalon alueella mm. kasvillisuudessa ja puustossa palon jälkeen tapahtuvia muutoksia sekä hiiltyneellä ja lahoavalla puulla elävää sienilajistoa. Osa polttoalasta ja viereisestä kontrollimetsästä on aidattu hirvieläinten laidunnustutkimusta varten.

Mitä ennallistaminen on?

Ennen suojelualueiden perustamista monet alueiden metsistä ja soista olivat talouskäytössä. Näin ollen vain osa niistä on nykyisin täysin luonnontilaisia. Ennallistaminen on luonnonsuojelualueilla toteutettava, yleensä kertaluontoinen toimenpide, jolla pyritään nopeuttamaan ihmisen muuttaman elinympäristön palautumista kohti luonnontilaa.

Controlled burning in Mikonsuonpalo

Mikonsuonpalo is located in the Natura 2000 site of Oulanka (29,390 ha), which is an important complex of old forests and mires. In the future, Mikonsuonpalo will be merged as part of the Oulanka National Park.

In 2006, Metsähallitus Natural Heritage Services burned seven hectares of former commercially managed forest in Mikonsuonpalo. Preparations for the burning started in the late winter by felling part of the trees to dry on the top of the snow. The area to be burned was surrounded by a treeless firebreak. Holes for extinguishing water were dug with an excavator into the nearby mire. The restoration burning was carried out at the end of July.

The burning was successful, creating a mosaic-like burn mark. The burning generated plenty of burnt, charring, and gradually decaying wood material needed by many endangered species to survive. The burning of the humus layer enables the development of deciduous tree stand, thereby diversifying the forest structure.

The Muhos Research Unit of Finnish Forest Research Institute studies the changes taking place in the flora and the tree stand after the fire and the polypore species living on charred and decaying wood. Part of the burnt area and of the adjacent control forest is fenced for a study on reindeer and elk grazing.

Nature restoration – What does it mean?

Before establishing nature conservation areas many of the forests and mires in these areas were commercially used; hence, today, only a part of them is in fully natural state. Habitat restoration measures are nature management in nature conservation areas. Most of the measures only need to be carried out once. The aim of the restoration is to speed up the reversion of habitats to their natural state.

Juha Siekkinen, Sanna-Kaisa Rautio

Vihreä vyöhyke Life
Green Belt Life

Vihreä vyöhyke Life on Euroopan Unionin LIFE Luonto -rahaston tukema projekti (2004–2008). Metsähallituksen luontopalvelut ennallisti hankkeessa metsiä ja soita Natura 2000 -luonnonsuojelualueilla. Lisäksi hankkeessa tuotettiin tietoa ennallistamisesta ja luonnonsuojelusta.

Green Belt Life project (2004–2008) is funded by EU's LIFE Nature fund. In the project Metsähallitus Natural Heritage Services restored forests and mires of Natura 2000 conservation areas. In addition information about restoration measures and nature conservation was produced.

    METLA  Kainuu

Figure 6.6
Public information board (left) describing prescribed burning for the conservation of biodiversity in Oulanka National Park, northern Finland. Oulanka covers 270 square kilometres and was declared a PANPark back in 2002. More recently, small plots, such as the one shown on the right, have been set aside for trial fires. These sites are carefully monitored to establish evidence of biodiversity recovery. Photographs by P. Hobson.

Figure 6.6 (*cont.*)

- Connectedness and connectivity
- Near-natural dynamic processes
- Stand structural integrity
- Landscape heterogeneity.

Applying a suite of strategies at multiple scales is designed to provide the right conditions for a wide range of species at least in some parts of the landscape – a principle referred to as risk-spreading.

A number of trans-international organisations such as the Taiga Rescue Network have contributed to a regional framework for the sustainable management of the boreal zone. The guidelines are centred on principles of landscape ecology that focus on integrative matrix management, avoiding compartmentalisation of the landscape. To complement these large spatial initiatives a more pluralistic, adaptive strategy is proposed to underpin the policy framework. The process of communicating and implementing targets for sustaining biodiversity can be summarised in the following steps:

Step 1. Establishing a frame of reference. Provide a detailed description of the natural disturbance dynamics and patterns for the different types of forest within the boreal zone. Include a record of the landscape's historical state and factors that have shaped the character of the forest over time.

Step 2. Ecological/biological indicators of the system. Identify a predictor-set of species and ecosystem processes that are indicators of 'free-willed' systems and are likely to be lost to human-induced changes in forest structure. Features likely to be included are altered fire frequency, hydrological regimes, predator–prey relationships, invasive species, area-sensitive species and browsing by super-abundant wild herbivores.

Step 3. Communication and coordination between science, policy and practice. This is a complex process that in the past has operated across a segregated system of operations. An ecosystem approach is advocated that involves all three domains in the decision-making process. It also requires adaptive systems that reflect patterns of uncertainty, change and appropriately scaled cycles.

Step 4. Evaluating and monitoring condition and change. The current process of assessing favourable conditions for biodiversity is fundamentally weak as it puts the emphasis on maintaining the components rather than the processes of a system. This approach favours the status quo rather than a dynamic, changing and evolving ecosystem. New criteria are called for which reflect more accurately the function and resilience of a boreal forest.

Step 5. Natural banking policy. A long-term strategy that operates to the precautionary principle. In the face of environmental uncertainty and the limitations of current scientific evidence it is prudent to protect or bank parcels of natural habitat to secure ecosystem services and sustainability of the wider landscape. Retaining large parcels of self-willed land helps to secure the biodiversity of a region and this in turn promotes ecosystem resilience and function.

Fire as a management tool in the landscape

As described in Chapter 2, we humans have been using fire throughout our long history to modify the landscape, sometimes for good and sometimes at the cost of entire ecosystems or the loss of species. At various times during this journey fire has been embraced and at other times vilified – see the excellent books by Steve Pyne listed in Further Reading for more details. Most of us are increasingly removed from direct contact with fire, particularly in the landscape, and fire can appear to be a very rough and capricious tool with which to manipulate our wild surroundings. This comes home particularly when we hear environmentalists advocating fire while seeing in the media pictures of destroyed houses and the newly homeless. As described in Chapter 5, in fire-prone forests, fire is an ecological necessity and yet socially and economically it can be disastrous when it burns homes and communities and destroys timber destined for the mill.

Nevertheless, fire can still be a useful tool in managing the landscape. Naturally occurring fires can be allowed to burn or (used since the mid 1960s) fires can be deliberately lit as and when they are needed – prescribed fires. It is a relatively cheap tool to use (despite the equipment needed – see Chapter 7) and it has a number of direct benefits. In forestry it is particularly useful in removing branches and other debris (slash) left over from tree felling which could otherwise harbour fungal diseases and get in the way of any tree planting that is done. This removal also has the advantage of reducing wildfire risk. Fire can also be used to remove the build-up of fuel left by years of fire suppression but has to be done carefully to prevent the fire becoming too intense to control. This usually involves burning under cloudy, cool or moist conditions but this may produce fires so low in intensity that much of the fuel is left unburnt. Fire has a number of other uses within forest management, including the reduction of damage by insects and other pests and diseases. For example, in eastern Canada the red pine cone beetle (*Conophthorus resinosa*) is a serious pest in red pine (*Pinus resinosa*) seed orchards – trees grown for their seeds. Since the beetles spend the winter on the ground inside hollowed-out shoots, and since the trees have a thick fire-resistant bark, fires in spring before the beetles emerge can be used to control the pest. In southern USA fire has been used to control brown-spot disease (caused by the fungus *Scirrhia acicola*) in longleaf pine forests (*Pinus palustris*).

One of the main problems of using fire as a management tool is the perceived risks to property and lives; this is considered below.

Fire and soils

There have also long been concerns that fire leads to impoverishment of the soil. On the one hand, fires are useful for releasing the nutrients stored in dead branches and soil organic matter. This surge of nutrients can be useful to newly growing plants although the boost in growth of understorey plants can provide serious competition for tree seedlings. The nutrients are useful providing plants can absorb them before they are leached away by rain or melting snow. A study by Grier (1975) in Washington State, USA found that following an autumn fire more than three-quarters of the magnesium, potassium and sodium were lost by water from the spring snowmelt percolating into the ground.

On the other hand, there are concerns that the fires can be directly responsible for an overall loss of nutrients from the forest (Fig. 6.7). Burning releases large

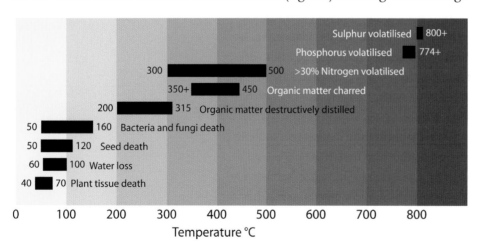

Figure 6.7
Temperatures at which various fire effects occur, from plant death which begins at between 40–70 °C to loss of sulphur at temperatures above 800 °C by volatilisation or vaporisation from a solid state to a gas. Data from Neary *et al.* (2005).

quantities of carbon and nutrients to the atmosphere either in particles or by vaporising the nutrients into gases which are blown away. This affects air quality and has been suggested as leaving the forest impoverished of essential nutrients.

Nitrogen (N) is the most commonly limiting nutrient in forests (Thomas & Packham 2007). Once fire temperatures reach around 300 °C nitrogen starts to be lost. In North America it has been estimated that nitrogen loss is 10 kg of N per hectare/ha in gentle understorey fires in pine stands rising to over 500 kg N/ha in post-logging slash fires in Douglas-fir stands (Carter & Foster 2004). Nitrogen is returned naturally to the forest through rain at a rate of 1 kg of N per hectare per year (ha^{-1}/y^{-1}) although pollution often brings this to above 5–10 kg N ha/y (Thomas & Packham 2007). Nitrogen-fixing plants and bacteria can also return nitrogen to the forest soil at rates of 5 to 150 kg N ha/y annually for several years or decades. Other nutrients (and carbon) are also lost in a fire, starting with phosphorus when the fire temperature gets above 774 °C (Fig. 6.7) and progressively losing more nutrients as fire temperature rises until calcium is lost above 1240 °C. In most unmanaged forests, where the fire frequency has not changed much by human interference, it would seem that the nitrogen and other nutrients lost in a fire are replenished before the next fire: these forests have been honed by thousands or millions of years of more or less fire and so have developed a quasi-stable relationship between burning and soil nutrients (see Certini 2005 for more details). In forests used for timber production, where the crop is cut and the debris burnt on a more frequent time scale than naturally occurring fire, it may well be that nutrients will be progressively lost and fertilisation is needed to maintain productivity. Moreover, there is evidence that as N is lost, the rate of decomposition of litter slows, locking up the remaining N and further decreasing its availability to plants (Hernández & Hobbie 2008). Decomposition is also slowed by the heat of the fire forming 'pyromorphic humus' or 'black carbon' which is remarkably resistant to chemical and physical degradation (González-Pérez *et al.* 2004). Black carbon typically makes up 1–6% of soil organic matter but can reach 35% in the old human-produced Terra Preta soils of the Amazonian Basin (see Thomas & Packham 2007). The problem of nutrient loss is exacerbated if the logging debris (slash) is piled or windrowed before burning. The long, intense fires that develop can sterilise the soil beneath, especially important in that this also locally kills the mycorrhizal fungi (Korb *et al.* 2004) and plants colonising these areas grow less well since they are less efficient at extracting nutrients from the soil without the fungi.

The wildland–urban interface (WUI)

Around the world, urban areas continue to expand into fire-prone wildland areas, creating the wildland–urban (or urban–rural) interface. This is defined as the area where buildings, especially homes, extend into wildland vegetation, and where they are therefore exposed to the risk of wildland fires. Given the size of major cities in most fire-prone countries, this interface might seem relatively unimportant (except, of course, to those who live there!). But, to take an example, in the lower 48 states of the USA, Radeloff *et al.* (2005) calculated that the WUI covered over 700 000 km^2 or 9% of land area and contained 44.8 million houses (39% of all houses). Moreover, in 2000 over 3 million ha of the USA burnt and 861 structures and 16 lives were lost. And between 1991 and 2000, wildfire caused at least $3.2 billion in damage to homes and other properties in the USA (see Zhang *et al.* 2008). This interface is therefore a growing problem in terms of protecting people and their property at an increasingly burdensome cost.

Exactly where the boundaries of this interface lie depends upon the need of the manager. At one end of the scale, risk mitigation assessment within a community might be based on aerial photographs with a resolution of 15 cm or so whereas on the large scale, concerned with large fires moving across the landscape, the interface has been taken as any houses within 2–15 km of undeveloped areas such as National Parks (for example, see Kaval *et al.* 2007). Gill & Moore (1998) report that in the Sydney fires of 1993–1994, most houses that were damaged were on the immediate edge of the fire but some houses up to 17 blocks from the major fire areas were affected. Houses at the edge of a fire ignite primarily from direct flame contact or by the heat radiating from the fire igniting curtains or soft furnishings *inside* the house. Those further away are more likely to catch fire from burning debris carried aloft and dropped some distance from the fire (the spot fires described in Chapter 4). These burning embers can be aided by the large amount of unburnt material such as leaves that are also carried before the fire. As this builds up on and around buildings it acts as tinder for the burning embers to develop into something larger. It is not unusual after a WUI fire to find burnt spots in wooden decks some distance from a fire front where embers and blown debris started small fires which fortunately did not build into anything larger. Homeowners can of course help protect their properties – this is described below.

After a fire has burnt through a community there are a number of problems beyond the loss of possessions and lives. For example, in dry areas, where WUI fires are perhaps commonest, there are often issues of soil erosion before vegetation recovers. This is especially a problem in areas with a Mediterranean climate where summer fires are normally followed by heavy autumn rains, potentially causing heavy soil erosion before the natural vegetation can recover. A short-term emergency solution is to spread seeds of fast-growing plants immediately after a fire, usually in a mulch of chopped straw or similar (to retain water) with added fertiliser to speed growth. The seeds used are usually of grasses and clover since the grasses are quick growing and the clover adds nitrogen to soil to help the grass. The resulting vegetation cover over the first year or two may not be extensive (typically < 15% cover of the ground) but the roots below ground can reduce soil loss by up to 10 tonnes per hectare each year, although it is usually substantially less. In the short term, the added vegetation tends to displace the natural vegetation, but usually by 4–5 years after sowing the native flora reasserts itself and the introduced plants fade away. This is not always the case, and a compromise is needed between damaging the native vegetation and preventing erosion. For example, around the Mediterranean Sea annuals make up a very large part of the staggeringly high biodiversity and sown grasses can compete against these and drastically reduce recovery. Fortunately, it is becoming increasingly apparent, that the mulch of straw is just as good by itself (see Fig. 6.8). Following the Fox Creek Fire in north-west Montana in July 2002, Groen & Woods (2008) found that applying a straw mulch application at the rate of 2.24 tonnes per hectare reduced erosion by 87%.

The role of prescribed burning in wildland–urban interface areas

Fire suppression was successful for years in removing fire from near people. But as fuel builds up and we appreciate the ecological necessity of fire, there is a strong imperative to reintroduce fire, to allow fires to burn in a controlled way. Unfortunately, it is more difficult now that cities and towns sprawl into wildland areas and private houses are being increasingly built on steep slopes and within narrow canyons, creating an enormous hazard to using fire to manage the land.

One of the main problems of using fire as a management tool is the risk of fires escaping with all the ensuing problems, not least of which is the threat of being sued by another landowner whose land is inadvertently burnt. Fear of liability is often the most significant barrier preventing private landowners from using fire as a management tool, and fear about safety and control is a major concern of residents, making prescribed fire an unpopular management method (McGee 2007, Vining & Merrick 2008). But this is not always the case. A survey in Colorado by Kaval *et al.* (2007) found that 85% of people living in fire-prone areas approved of public land managers using prescribed fire and, moreover, were willing to pay around $800 a year for this to be done. Their willingness may be because these homeowners used to burn around their own properties until this became illegal and so undoubtedly were more comfortable with fire than those homeowners with no experience. In Australia Gill & Moore (1998) reported that although feelings about prescribed fire were mixed, prescribed burning involved even back then a substantial volunteer force of about 250 000 people.

The smoke produced by prescribed fires is a perennial issue. It is toxic stuff, containing both particulate matter (implicated in respiratory problems, see Chapter 3) and gaseous compounds such as carbon monoxide, formaldehyde, benzene, nitrogen dioxide and ozone. Past experience shows that it can cause significant health problems. For example, the uncontrolled forest fires burning in Indonesia in

(a)

(b)

Figure 6.8
(a) A house on Inverness Ridge at Point Reyes, California that escaped the Vision Fire of 1995, and (b) the remains of a house that did not escape, once the movable debris had been cleared. The fire started on 3 October 1995 from the remains of a campfire and burnt for four days, covering 5000 ha and destroying 45 houses.
In (b) straw can be seen that was spread to reduce erosion. This worked by reducing the impact of raindrops on the exposed soil (which otherwise tend to break up the soil structure reducing infiltration of the water), slowing the flow of water over the ground and holding moisture for plant recovery.

1997 affected some 12 million people, resulting in 1.8 million cases of bronchial asthma, bronchitis and ARI (acute respiratory infection), according to the World Health Organization. Smoke from wildfires often breaches clean air regulations but is seen as an act of God (unpreventable) while that from prescribed fires is some-one's fault. This can be a huge impediment to using fire as a tool.

Another problem of prescribed burning is in the compromises needed in trying to meet the many goals of the fires. For example, prescribed burning in the bush

Box 6.4. Protecting your home: managing your landscape to reduce fire risk

There are a number of actions that can be taken to reduce the risk of your house burning. These include such basics as using fire-proof materials to build the house. But even when you buy a house that is flammable, there are a number of changes that can be made to the surroundings of your house to decrease risk during a fire. The top picture shows some of the landscaping problems that can assist a fire in damaging a home; lots of flammable materials near to the house. The picture at the bottom shows how this can be improved.

What follows is an extract from *Is Your Home Protected from Wildfire Disaster*, published by the US Firewise Communities programme which explains the zones shown in the bottom picture.[1]

Zone 1: Establish a well-irrigated area around your home. In a low hazard area, it should extend a minimum of 30 feet [9 m] from your home on all sides. As your hazard risk increases, a clearance of between 50 and 100 feet [15–30 m] or more may be necessary, especially on any downhill sides of the lot. Plantings should be limited to carefully spaced indigenous species.

Zone 2: Place low-growing plants, shrubs and carefully spaced trees in this area. Maintain a reduced amount of vegetation. Your irrigation system should also extend into this area. Trees should be at least 10 feet [3 m] apart, and all dead or dying limbs should be trimmed. For trees taller than 18 feet [5.5 m], prune lower branches within 6 feet [2 m] of the ground. No tree limbs should come within 10 feet of your home.

Zone 3: This furthest zone from your home is a slightly modified natural area. Thin selected trees and remove highly flammable vegetation such as dead or dying trees and shrubs. So how far should Zones 2 and 3 extend? Well, that depends upon your risk and your property's boundaries. In a low hazard area, these two zones should extend another 20 feet or so beyond the 30 feet [6–9 m] in Zone 1. This creates a modified landscape of over 50 feet [15 m] total. In a moderate hazard area, these two zones should extend at least another 50 feet beyond the 50 feet in Zone 1. This would create a modified landscape of over 100 feet [30 m] total. In a high hazard area, these two zones should extend at least another 100 feet beyond the 100 feet in Zone 1. This would create a modified landscape of over 200 feet [60 m] total.

The importance of maintenance
Once you have created your home's survivable space, you must maintain it or risk losing the benefit of its protection.

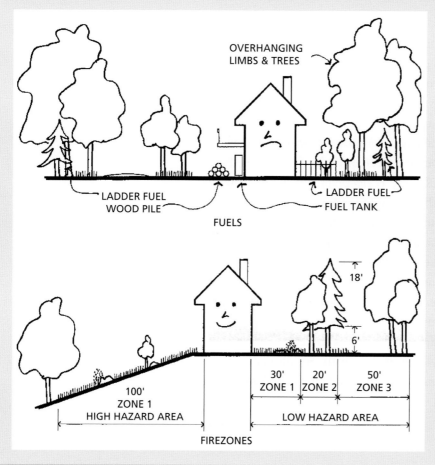

1. Firewise point out that this information is designed to provide homeowners with guidance on ways to retrofit and build homes to reduce losses from wildfire damage. It contains suggestions and recommendations based on professional judgement, experience and research and is intended to serve only as a guide. The authors, contributors and publisher disclaim all warranties and guarantees with respect to the information in the document and assume no liability or responsibility with respect to the information.

around Sydney is only effective when enough fuel has built up to carry a low-intensity fire, which takes around 5 years since the previous fire; but wildfires can consume fuels only about 2 years old (Conroy 1996). To exclude wildfire is then extremely difficult, if not impossible, and costly. So prescribed fires can only aim to reduce fuels to a level where fire intensities are limited and fire suppression is likely to be effective. Further problems, however, come from conservation of the natural biota, including a notable plant species of the area, the heath-leaved Banksia (*Banksia ericifolia*). This can be eliminated by fires as frequent as every 5 years, but if prescribed fires are reduced to every 10 years, potential fire intensities and losses of buildings would rise and fire would likely become more frequent anyway, threatening the plant as a result. Also, frequent low-intensity fires can reduce the structural diversity of an area, depleting the small mammal fauna, especially given the number of pet cats and dogs which can easily catch the fauna after a fire. The upshot of this scenario and many others around the world is that the role of fire needs careful modelling within a system where the unexpected (a dry year, an escaped campfire) can throw a spanner into the most carefully laid plans.

Part of the problem with fire in the wildland–urban interface (WUI), whether a prescribed or wildfire, is that fires in a mix of buildings, cultivated gardens and wild vegetation are remarkably complex due to mixed and varying fuels (from short grass to propane tanks), wind eddying around buildings, changes in moisture and humidity, the need to be sensitive to people's property and accessibility of driveways to fire trucks. These all create unique problems for firefighters. There is also a good deal of variation in the flammability of individual properties, including whether the roof and walls are wooden shingles, metal or tile/brick, whether there is a deck or other landscaping features (Ramsay *et al.* 1996). Certainly, a house with burnable vegetation and articles such as firewood removed from an area of 30 m radius is much less likely to be damaged by a wildfire. But it is important to remember that removing fuel by prescribed burning, breaking up fuels into smaller scale mosaics and creating a fuel-free 'defensible space' around each property is more likely to be proof against low-intensity fires – see Box 6.4. In the largest firestorms, which appear to be becoming more frequent in drier parts of the world such as south-east Australia and California, these measures. may have very little effect since, as discussed in Chapter 4, these fires are driven by weather not by fuel and are likely to burn everything in their path. So, no matter how carefully the wildland–urban interface is managed there will always be losses. It is not a problem that is going to go away.

7 | Fire suppression

Prior to the arrival of humans, fires came and went, the forest grew, burnt, and regrew. Fire was a systematic, natural process within many ecosystems. With the arrival of early humans, fire became more prevalent on the landscape (with the new fire starters), and the ecosystem adapted to the changing regime. It is only in the last 100 years or so that humans have found the need to systematically suppress fire and attempt to eliminate its destructive nature – essentially to tame it. Suppression activities began in earnest after World War II when aeroplanes, helicopters, smoke-jumpers and new firefighting strategies were introduced. Why have we attempted to remove this element of the ecosystem from its natural role? The answer is simple, of course – fire competes with us for natural resources, fire threatens and destroys our property and it can kill. In many parts of our world, the economy is dependent on the renewable resources found in the forest. Our lives and lifestyles also depend on homes, buildings, telecommunication towers and lines, pipelines and a host of other infrastructure elements in and around wildlands (natural or semi-natural forest) at risk of fire. The normal feeling is that fire cannot be allowed to threaten and disrupt our lives and economy. The existence and mission of fire suppression is thus based on the protection of human life, property and the resources upon which economies depend. In some areas of the planet, where there is a lack of human habitation and/or economic value, or where conservation aims at 'naturalness', there is no fire suppression, or the suppression activities are less aggressive. For the rest, however, fire suppression is the norm. While the techniques and technologies are as varied as the local cultures and ecosystems they protect, suppression follows a similar organisation. Wherever you are, fire suppression follows the same pattern of activities, and it all starts with finding the fires; but before that, you have to know where to look.

Preliminary steps – fire intelligence

Fire intelligence does not refer to how clever the fire is, rather it is the gathering of information on potential fire behaviour and occurrence. Modern fire-management organisations collect a plethora of data concerning the area under their care. The information can be categorised under three broad areas; weather, fuels and topography, and 'values'.

Box 7.1. Fire danger rating – the Canadian Fire Weather Index System

Fire danger rating is defined as: an assessment of the fixed and variable factors of the fire environment that determine the ease of ignition, rate of spread, difficulty of control and fire impact. The Canadian system of fire danger rating shown in Fig. 7.1 is much like any of the other systems and is itself used across Canada and by many other countries around the world. A brief description and discussion will provide insight as to how it works and why it is important for fire-management agencies.

The Canadian Forest Fire Danger Rating System comprises two major systems: the Fire Weather Index System described below and the Fire Behaviour Prediction System (see Box 7.2).

The Fire Weather Index System (FWI) was developed in the mid to late 1960s and first released in 1970. It provides standardised values that rate fire danger potential in a pine forest fuel type – no adjustment is made for fuel within the FWI System. The FWI System converts raw weather data (temperature, relative humidity, wind speed and 24-hour total precipitation), measured at solar noon into a set of three fuel moisture indexes, and three fire behaviour indices, all numerically based. All indices are open ended (there is no upper limit to the values) except for the Fine Fuel Moisture Code. The three fuel moisture indexes each reflect a different layer in a forested wildland fuel:

1. The Fine Fuel Moisture Code (FFMC) rates the moisture content of forest floor organic litter and other dead fine fuels. Since fine fuels are the primary carrier of a fire spread, the FFMC indicates the relative ease of ignition, and flammability of the fine fuels and the forest.
2. The Duff Moisture Code (DMC) rates the moisture content of the upper organic layers of the forest floor (in Canada referred to as duff). This index is valuable for estimating fuel consumption, and lightning ignition potential.
3. The Drought Code (DC) rates the moisture content of the deep organic layers of the forest floor. The DC is valuable for assessing longer-term drought, and indicates the potential for fires to burn deeply into the ground, complicating the mop-up (final extinguishing) of fires.

The three fire danger indices combine elements of the three fuel moisture codes to provide refined pictures of potential fire behaviour.

1. The Initial Spread Index (ISI) combines the FFMC with wind speed to provide a relative measure of potential rate of spread.
2. The Buildup Index (BUI) combines the DMC and the DC to provide a relative measure of the amount of available fuel for the fire.
3. The Fire Weather Index (FWI) combines the ISI and BUI to provide a general measure of fire danger, and an indication of potential fire intensity.

The values produced by this system are always relative – there are no units and the values can never be measured in the field; however fire managers build a 'feel' for the numbers and develop rules of thumb for how to incorporate the numbers into decision making. Other danger rating systems are obviously different; however there is always a lot of commonality between systems, each one suited to different climatic conditions. Fire danger rating values such as these form the backbone of a fire-management decision-making process, and serve as guides to a variety of decisions.

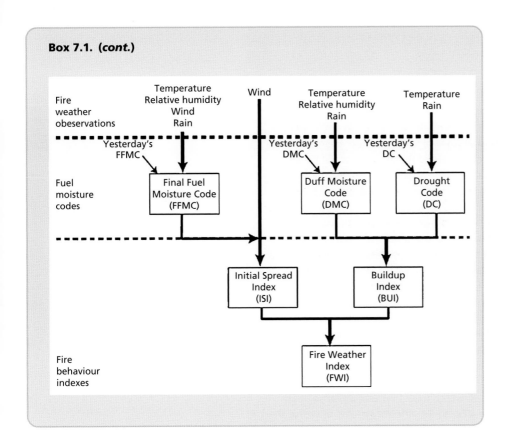

Box 7.1. (*cont.*)

Figure 7.1
Structure of the Canadian Forest Fire Weather Index System from Lawson & Armitage (2008). Canadian Forest Service (Natural Resources Canada), reproduced with permission.

Weather

Weather information is gathered for two reasons – first to estimate the fuel moisture conditions through mathematical 'moisture book-keeping' models and generally keep tabs on current weather parameters that influence fire behaviour, and second to anticipate what future weather events might hold in store for firefighters. The impact of both fuel moisture and wind on fire behaviour has been shown in Chapter 4. Fire intelligence aims to track weather patterns and fuel moisture conditions across regions so that all parts of the organisation are adequately prepared. Using a network of weather stations, maps of fuel moisture and fire danger indexes (see Box 7.1) are produced. In addition, networks of lightning detectors track lightning strikes as they occur. Adding all this information together, a fire manager builds an understanding of where and what the daily risks are, for both fire ignitions, and once started, how aggressively they might behave. As we will see this information is all-important in planning the days' activities.

Fuels and topography

While fuel types (and especially topography) do not change with the rapidity of the weather, fire managers still need to incorporate this information into daily planning.

Box 7.2. Fire behaviour prediction

While fire danger rating is broad scale and generalised, fire behaviour prediction is small scale and precise. It focuses on assessing the fuel complex characteristics, fuel moisture, wind speed and direction, topography and upper atmospheric conditions to predict: how fast a fire will spread (the rate of spread); how much fuel will be consumed; whether the crowns of the forest will be involved (a crown fire); and how much energy will be released or how intense the fire will be (which is related to the length of the flames). Fire behaviour prediction is used operationally in many ways, from assessing the potential behaviour and growth of a nascent fire, to forecasting where a large fire will be hours into the future.

Globally there are several systems of fire behaviour prediction. Each of these systems link fuel, moisture, weather and topography through a series of mathematical models to make predictions of fire behaviour. One of the most widely used fire behaviour models is the one developed by the USDA in Missoula under the leadership of Richard Rothermel – the BEHAVE system (Andrews 1986). THE BEHAVE system utilises the Rothermel fire spread equation to develop estimates of fire behaviour. Its great strength is that the equation requires fuel complexes to be described in physical terms – e.g. fuel density, surface area to volume ratio etc. This allows the model to be extended to virtually any fuel complex that can be described. Other systems of fire behaviour prediction (e.g. the Canadian Forest Fire Behaviour Prediction System) use only established, preset, discrete fuel types that the user must select from. This limits the range of application; however proponents would argue that the model itself works better over a wider range of fire behaviour conditions. The approaches to developing the different systems, as well as the behaviour prediction systems themselves, have inherent strengths and weaknesses, and there are strong proponents for each.

Interestingly no operational fire behaviour prediction system incorporates the influence of atmospheric stability into a prediction of fire behaviour (see Chapter 4). There are models that predict when atmospheric stability will affect fire behaviour, but these are separate and not integrated into the operational models. This underlines a statement made by Charlie Van Wagner of the Canadian Forest Service who said that fire behaviour prediction was an 'art and science'. While the science provides mathematical relationships that predict fire behaviour, it is the art of the fire behaviour analyst to know what fuel model best represents the fuel complex at hand, how the wind will be influenced by the local topography, when the atmospheric stability will override the surface weather influences, and where and what are the potential trigger points for extreme fire behaviour.

We have seen in Chapter 4 that different fuel types can have profound effects on fire behaviour, so it is important to have a clear understanding of where the troublesome fuels are and map out important topographical features. Like the weather, fuel and topography can be portrayed on a map, but what is more important is combining weather, fuels and topography into maps of potential fire behaviour. Fire behaviour prediction models have successfully been developed in Canada, the USA and Australia. Through these models, maps of potential rate of spread once a fire starts, fire intensity and a host of other fire behaviour parameters can be displayed (Fig. 7.2). This can prove invaluable when assessing the fire potential for the day, and for decision making during the day.

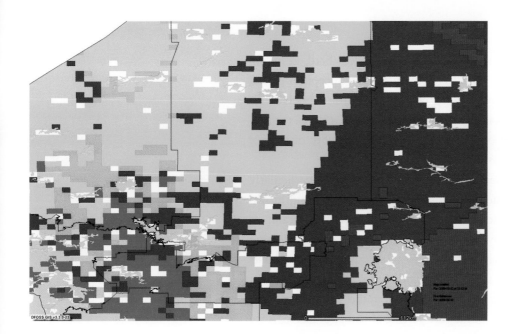

Figure 7.2
A map of potential head-fire intensity incorporating topographical effects, fuel type changes, fuel moisture conditions and weather. The pixilation is caused by the contributing elements averaged to a set cell size. Colour coding is graduated to reflect relative potential: green – low, blue – moderate, yellow – high, orange – very high, red – extreme. Map from northern Ontario, Canada.

Values

Values, in fire-management jargon, refer to those locations where life or property would be threatened by a fire. This can be anything from a collection of homes in the wildland–urban interface, to a trapper's shack in the wilderness, to a gas pipeline pumping station, to a microwave communication tower. While these locations might seem obvious to those who live in the area, fire managers responsible in some cases for vast expanses of land require up-to-date values information, to determine who and what might be threatened in the event of a fire. This is especially important when suppression action proves ineffective and people must be evacuated from harm's way.

Adding it all up – the fire intelligence function

Most fire-management organisations the world over employ some form of computerised fire-management information system. These systems track not only all the information about weather, lightning, potential fire behaviour, fire occurrence probabilities and values at risk, but can also display available suppression resources, including their location, status, capability etc. When a fire is reported, the potential fire behaviour along the likely path of the fire can be evaluated, appropriate suppression resources dispatched, public warnings issued, suppression progress tracked and fire growth predicted. Before a new day starts, plans, based on weather and fire-occurrence predictions, are formulated to ensure adequate firefighters and other suppression resources are well positioned to meet the coming threat. Fire intelligence forms a backbone of the fire-management organisation ensuring effective and efficient operations.

Step 1: fire detection

There is an old adage in fire suppression – 'find 'em fast, hit 'em hard, keep 'em small'. If the fires are found when they are small, in their early inception period before they get big or nasty, there is a much better chance of control: a much better chance, and at a much lower cost. This then has become the *raison d'etre* of the fire-detection business – early detection of fires increases the probability of control success and potentially reduces the cost of fire management.

We have seen in Chapter 4 that there are a variety of fire-starters, particularly lightning and humans. Fire-management organisations have evolved and grown over time with these causes, and implicitly design their fire detection programmes to meet the local conditions. Many factors affect the choice of detection method including population size, local values, fuels, topography and, of course, the commonest ignition sources. These same factors (and others) also have an impact on the design of many of the other suppression programme components (for example, attack base size and location). But we are getting ahead of ourselves – we should start with how fires are found and how that process has evolved over time.

In the beginning, fire detection was rudimentary at best, as was the fire suppression organisation. When fire suppression was first started in North America, forest rangers or wardens would patrol the backcountry on foot, horseback or canoe looking for fires. While they did find fires, the system was far from efficient, as fires could be burning for days or weeks, before being found, and once found, only the local patrol was typically available to fight the fire. Although forest rangers were out looking for fires, it has always been the people who live, work and travel through the wildlands that report the greatest number of fires. Clearly if people are around to start a number of the fires, they are around to report them. In the early days, reporting was difficult with limited methods of communication, but as technology evolved, communication methods improved to capitalise on this method of detection.

Current detection efforts are helped by a host of technological aids, but the detection programme is closely tied to its roots – public reporting of fire incidence still approaches 50–100% of fires depending on population density. Some fire-management agencies, capitalising on this, break their detection programme into two categories: one that seeks to support and enhance public reporting; and the second that specifically organises activities to seek out unreported fires. These two groups go by a number of different names depending on the organisation; unorganised and organised, indirect and direct, unplanned and planned, or passive and active. Whatever the name (and here we will use passive and active) the actions add up to the same groups.

Passive detection

Passive detection is just that – detection based on the random travels of the public over the area of interest, and providing people with simple contact methods to report fires. This method is surprisingly effective over a broad range of areas, even where population density is relatively low. Reports come not only from people in the area but also from planes flying over. The shortcoming of public reporting of forest fires is the quality and quantity of the information coming in. Fire suppression people want as much information about a fire as possible – not just where it approximately is, but:

- How big is the fire?
- What kind of fuel is it burning in?
- How big are the flames?
- What kind of access is there to the fire?

All this information helps in the subsequent decisions that must be made by the dispatcher. So while public reporting is an effective method, it has a number of limitations, and cannot be relied upon in many areas.

Organised detection

Organised detection is systematic detection designed by fire-management organisations to locate fires where and when they occur. While there are a number of new technologies that show promise, detecting fires is currently primarily done either through fixed detection networks (towers with observers) or through aerial detection (planes flying set patterns looking for fires).[1] In either case designing the detection programme depends on a number of factors including:

- **Value of the resource protected.** Low value locations can perhaps tolerate more fire, and thus the investment in intense detection is unwarranted (more information on investment and economics later).

- **Visibility.** The ability of an observer to see the smoke from a fire. This is affected by the size of the fire (the amount of smoke), the atmospheric conditions at the time of observation (clear versus hazy) and topography (the observers' position relative to intervening hills etc.).

- **Probability of a fire occurring.** If a fire is an extremely rare event, then detection might be less rigorous. Why look for something aggressively (and spend time and money doing it) when it only happens once every 20 years? We have seen in Chapter 4 that a number of factors contribute to fire incidence in a region.

- **Expectations of fire behaviour.** When the weather conditions would not support rapid fire spread, detection can be reduced, because although it will take longer to find the fires, they will still be small and manageable. Conversely when extreme fire behaviour is expected, detection must be intensified to ensure fires stay small.

---

1. Actually some jurisdictions, at least in Canada, still use foot patrols, but these patrols have a host of other duties to perform, and are a relatively inefficient method of fire detection.

- **The potential for fire spread.** Fires occurring on small islands on a large lake, as long as there are no values on those islands, might not require intensive detection (or suppression) because they cannot spread off the islands to become big.

- **Coverage by passive detection.** Passive detection is not, of course, an all-or-nothing option. As the density of coverage of passive detection dwindles due to reduced population density, etc., active detection can be increased to provide a consistent coverage.

Detection can be made easier by tools that help predict where a fire might start. Chief amongst these is the lightning detector (Fig. 7.3). The National Lightning Detection Network currently used in the contiguous US states (which also covers Alaska and lower Canada), can pinpoint lightning strikes to within 500 m. Lightning location works by detecting the low-frequency electromagnetic radiation emitted by lightning. Efficiencies of up to 95% detection of cloud-to-ground strikes (see Chapter 4) are claimed for modern systems.

Fixed detection

Fixed detection takes the form of fire towers (Fig. 7.4a) or lookouts strategically positioned across a landscape. Fires are located by two or more towers sighting the

Figure 7.3
A map of lightning strikes recorded in real time. Colours refer to how long ago the strike occurred, while sign (+/o) refers to the polarity of the strike (o is negative). The inset photograph shows a lightning detector (courtesy of Mike Franchuk, Ontario Ministry of Natural Resources, copyright 2009 Queens Printer Ontario).

(a)

(b)

2003/06/26

Figure 7.4
(a) The Yellow Head fire tower in Alberta, Canada. Fire towers range from towers like this to remote helicopter access cabins on mountain tops. (b) View from inside fire tower showing the simple but effective sighting tool (an alidade). Viewing a fire from two fire towers through the alidade will triangulate the fire location. Both courtesy of Alberta Sustainable Resource Development.

fire, and calling in bearings from the tower locations (Fig. 7.4b). Knowing the tower locations and bearings to the fire from two or more towers, it is possible to triangulate the fire's precise location.

Visibility from any tower is dependent on the environmental factors listed prior, but even in excellent atmospheric conditions topography plays a part. Intervening hills and ridges produce screened and blind areas (areas not directly visible but from which smoke may be seen) making pinpointing the source difficult. Fires occurring in these areas are not directly visible from the tower, and, depending on the degree of screening, can remain unreported for some time. This problem underlines the strategic importance of tower location selection; however, sometimes the best site for an area can still only produce 30% directly visible land. That being said, towers can produce excellent results, providing continuous monitoring of areas up to 80 km (50 miles) radius around a tower.

Up until recently towers have been staffed by people. Men and women who would stay in a mountain-top cabin, or climb up a tower and spend their summer days scanning the horizon searching for smoke. Recent years have seen ageing fire towers decommissioned due to safety concerns, and have left fire-management organisations scrambling for technological solutions. Emerging from this are tower-based video camera networks which combine thermal infra-red (heat) scanning technology with visible light and a computerised automated alarm system. The result is networks of cameras on small towers linked into a central computer in a dispatch centre. The cameras feed information into the computer that looks for smoke and heat sources, and warns when a discovery has been made. This technology is already in place at some locations around the world and can be effective up to 50 km (30 miles) around a single tower. Depending on the sophistication of the computer analysis system and the terrain, this does not always require two tower sightings of the fire to triangulate a location.

Aerial detection

Aerial detection began almost with the birth of flight as fire managers quickly saw the potential for this new technology. Aerial detection involves flying aircraft along set lines across a landscape searching for fires (Fig. 7.5). Planes selected for this duty are usually high fixed-wing aircraft (rather than helicopters) that afford good speed and good visibility for both pilot and observer, at a reasonable cost. Each day detection routes are planned out to address the daily detection needs. Planes fly at 600–1500 m above ground level, launched at specific times of the day to maximise detection probability. Because planes fly over an area, and leave, they do not provide the advantage of continuous coverage, and thus fires that puff up smoke intermittently can be missed. Aerial detection tactics require flight plans to be located where fires are expected, and timed so that fires are active when the plane flies over. This can mean waiting until mid-afternoon when fires are most active before flying the planes.

In times past, detection patrols required a great dependence on maps to determine the precise location of the aircraft and the fire. With the advent of Global Positioning System (GPS) satellites, precise locations can be quickly transmitted and resources dispatched with greater confidence than ever before.

Aerial detection holds a number of advantages over fixed detection networks:

- It is flexible. As visibility conditions change, flight lines can be tightened up to reduce the distance observers have to search. If a lightning storm passes through, detection can follow the storm, essentially searching where and when fires are expected. Blind spots are minimised with aircraft flying directly over screened or blind areas.

- Excellent information can be gathered. Trained observers in the aircraft can provide a wealth of information to dispatchers concerning the fire behaviour (current and potential), and make recommendations on suppression resource needs.

- It is adaptable. If the weather conditions do not warrant it, aerial detection does not fly. Observers, pilots and planes can do other work at a central facility.

Figure 7.5
A small fire as seen from the cockpit of a fire detection aircraft. Weather conditions can make the precise location of the fire difficult to determine. Ontario Ministry of Natural Resources, copyright 2007 Queens Printer Ontario.

- It is comparably inexpensive when compared with the cost of a manned tower system which must include not only wages, but the costs of service (keeping the staff supplied with food etc.) and upkeep (physical maintenance).

Aerial detection also has its shortcomings; it is said (somewhat tongue-in-cheek) that a fire never appears until after the plane goes by; a reflection of the lack of continuous coverage by detection aircraft. Also, weather conditions can prevent aircraft from flying, invariably just when they are needed.

Waiting in the wings is the possibility of UAVs – Unmanned Aerial Vehicles. These small aircraft (with a 2+ m wingspan) can carry the camera technology currently being used in the new detection towers, with all the advantages of regular aerial detection. This could increase detection sensitivity and reduce overall costs.

Satellite technology has been considered for many years as holding great potential for fire detection, but it has not, as yet, lived up to expectations. A number of problems plague satellite detection of fires, including sensitivity (very small fires are hard to find), cloud cover obscuring regions, infrequent coverage (may be twice per day) and cost (a dedicated satellite for fire detection would be very costly). Satellites are, however, proving useful for finding and mapping larger fires in remote areas where fire management is limited to monitoring fires and detection is not actively practised.

The debate between proponents of both fixed (tower) and aerial-based detection systems continues, each with studies to back up their position that one is better than the other. It is, however, probable that a combination of fixed detection observing high-value areas, public reporting where good coverage is possible and aerial detection where other systems prove too expensive or not workable is the best compromise.

Step 2: dispatch

Dispatch is the act of sending resources to a fire. Seems simple enough and really it is. To work effectively, however, dispatch is dependent on a deep understanding of the interplay of fire behaviour and fire suppression. The goal is to send just enough resources to a fire to ensure suppression success. Too few resources and the fire can escape, cause excessive damage and require a lot more effort in the end to extinguish. Too many resources and you could find yourself short of resources if another fire (or fires) occur, and a hefty unnecessary expense in borrowing equipment from elsewhere. The job requires that the individual understands fire behaviour, and is able to match suppression resources to the expected fire behaviour and expected size of the fire by the time they arrive. It is no wonder that dispatchers are always looking for as much information about a fire as they can get, before they

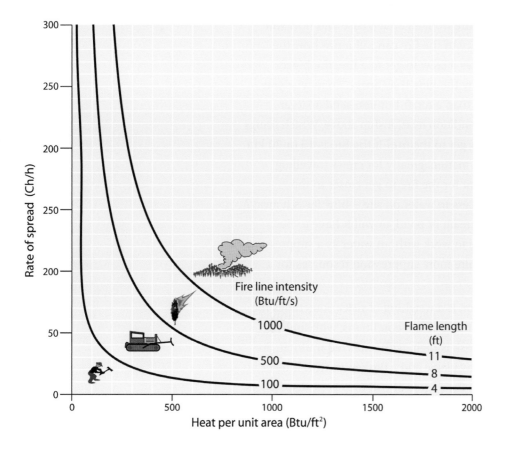

Figure 7.6
The Fire Characteristics Chart redrawn from Andrews & Rothermel (1982). Isolines of fire intensity delineate approximate control limits for different suppression tactics. The graph has been nicknamed the Hauling Chart as the different areas between the curves detail what resources must be hauled in for the job (or as in the most extreme case that all must be hauled out of the way). Note that 1 Btu/ft/s = 3.5 kW/m (see Temperature and energy, Chapter 3) so the three lines correspond to a fire intensity of 345, 1725 and 3450 kW/m, respectively, with flame heights of 1.2, 2.4 and 3.3 m, respectively. However the cut-off values are not absolute and metric versions of the chart produced later use fire-intensity isoline values of 500, 2000 and 4000 kW/m. The rate of spread is given in chains per hour (Ch/h). A chain is 66 feet or 20.1 m, so the highest rate of spread shown (300 Ch/h) is equivalent to 100.8 m/min.

make a decision. Clearly, more information reduces the uncertainty about the suppression force required to meet the firefighting job at hand.

What exactly is dispatched depends on the type of suppression resources available, and these are as varied as the agencies that fight fire. Firefighting resources can range from a gathered-up group of local people, to elite professional wildland firefighters parachuting in, rappelling from a helicopter or just climbing out of a vehicle, backed up by specially designed water-bombing aircraft (more details on these later). When selecting resource mixes to send to a fire, the dispatcher must consider four things:

1. **Fire condition and situation.** The intensity and size of the fire will dictate how effective the resources will be in fighting the fire. Chapter 4 describes the concept of fire intensity, and here is a practical application of the concept (Fig. 7.6). Firefighters on the ground must build a fire line (a non-flammable barrier – see Box 7.5 later in this chapter) around the perimeter of the fire faster than the fire grows – preferably a lot faster. A more intense, faster-growing fire will increase its perimeter at a quicker rate. Also, everything else being equal, the perimeter of a larger fire increases at a faster rate than a smaller one (for a circle, as diameter doubles, the perimeter increases by 3.1 times – $2\pi r$). Thus, more resources need to be sent to larger fires (not just because they are bigger, but because they grow faster) and to more intense fires.

2. **Location and access.** How far is the fire from an attack base and how will the firefighters get to the fire? Distance to the fire can influence how the resources get there. The longer it takes a crew to get to a fire, the bigger (and possibly worse) the fire will be by the time they arrive. So maybe you send the firefighters via truck rather than aircraft but you send a larger suppression force, knowing the fire condition relationship above. Which brings us to access – how does the fire crew get to the fire? If the fire is 15 or 20 km from the nearest road, it complicates matters greatly for crews travelling by truck, unless they are going to wait for the fire to get closer to the road. A second aspect of access is access to water on site – how far is it away? Can water be pumped from a local stream or pond, or will it have to be transported in, or will the crews have to work without water as a suppression tool? Fires fought without water require more suppression resources for a longer time.

3. **Values at risk.** Every dispatch decision is a weighing of risks. Sometimes the risk is very low – the fire is small, burning in wet conditions, with limited spread potential. But sometimes the risk is high, with expected multiple fire starts, limited suppression resources available and a fire report with little or no information other than a location. As the degree of risk increases, dispatchers consider more and more the values that are potentially threatened by a fire, and the risk of suppression failure. Given two simultaneous ignitions, in similar (moderate to high) fire behaviour conditions (which would normally result in a similar suppression success probability given equal resource dispatches), the dispatcher will usually provide a heavier dispatch to the fire threatening a community, or other substantial value, because the impact of a suppression failure is much worse. In all dispatch decisions, considerations must be given to the 'what-if scenario'. What if they fail to contain it? What might happen?

4. **What is available?** This may seem obvious (if there is no helicopter available you cannot send one) but it is not always so. Suppression resources are usually made up of a mix of different types with different characteristics. You may have different types of crews (different sizes and specialities), different types of transportation (trucks, helicopters, fixed-wing aircraft), different types of aerial firefighting support (helicopters with buckets, skimmer air tankers, land-based retardant air tankers – see below) all scattered across a landscape. Each is going to take a longer or shorter time to get to the fire; each is more or less suited to the particular fire situation; and any one of these might be needed for the next fire you do not yet know about. If the reported fire is off in a far corner of your area of responsibility, how much do you dedicate to that fire, drawing off the few valuable suppression resources you have, knowing they could not return if another fire started near a high-value site?

With all these conflicting pressures floating around in the dispatcher's head, one might think they could become paralysed with indecision, but that is simply not possible. Every day, with every fire report, they weigh the risks and send out the suppression resources.

Resources for fire suppression

In the last step we had a window into the life of a fire dispatcher, but here we should take a brief step back and look at the kinds of resources available for fire suppression, and some of their strengths and weaknesses. Firefighting resources can be broadly grouped into two simple categories – ground based and aerial based.

Ground-based resources

Ground-based firefighting resources can be further subdivided into people and machines.

People Firefighting crews come in all kinds of sizes – from small three-person 'Initial Attack' crews to eight-person crews to 20-person crews, and all sizes in between. Different fire-management agencies design their crews for the local situations and firefighting standard operations.

In Ontario, Canada, four-person crews are used that can be augmented with extra firefighters as burning conditions and operational requirements warrant. Augmenting crews is possible when crews are not too specialised: crew members have basic firefighting training and skills, but they are not required to parachute out of an aeroplane or rappel (slide down a rope) from a helicopter. In the case of Ontario, the fire crews are able to pump water from the myriad of lakes and rivers scattered across the Ontario landscape; consequently the crew size is designed around the construction of a fire line with portable pumps and hose. In western North America crew sizes are often larger (20); with the absence of water sources, fire crews build fire lines with shovels, axes and pulaskies (a unique firefighting tool similar to an axe but with a double-sided head – one side axe, one side hoe). This is slow work and crew sizes must be larger to achieve reasonable construction rates (Fig. 7.7). In this case fire-line production is directly related to the number of firefighters – more firefighters with shovels equals more fire line. Crews using a water pump and hose can only build a fire line at a set maximum rate – the water only comes out so fast.

Thus we see the primary difference between fire crews is the tools they use. The tools and the suppression methods they use define the crew size and organisation. While small crews can build a fire line effectively with pumps and water, when there is no water, small crews would be ineffectual as their rate of fire-line construction would be too slow to effectively challenge the fire. Beyond the tools, firefighters are

Figure 7.7
A US hotshot crew builds a handline providing a physical break of ground fuels. Courtesy of the US Department of the Interior.

often measured by the transportation methods they use to get to the fire. Across the USA, crews come in all sizes and shapes, depending on the region and agency they hail from, but most widely recognised are the smokejumpers. Smokejumpers are specialised wildland firefighters who parachute into remote areas to suppress wildland fires. While deployment to a fire site via parachute introduces a whole new set of risks for the firefighters beyond just fighting the fire, the technique can be particularly effective when the fire is located in a remote area, a long distance from the responding fire base. In this situation the fixed-wing aircraft used for smokejumping is both faster and less expensive than helicopters. Crews arrive at the fire faster and therefore have a better chance of suppression success than if another transportation method was used giving the fire more time to grow. Parachuting firefighters onto fires is not unique to the USA; they are also employed extensively in Russia, and to a minor extent in Canada. There are two other primary methods for transportation to the fire and probably the easiest and most obvious way is by fire truck or ground transportation. Ground transportation is most versatile in that trucks come in all shapes and sizes, and can carry a wide range of suppression resources (including water) that are well suited to the conditions into which they are deployed. However, vehicles require roads, and firefighters can be slow to arrive at a fire site, depending on the nature of the road network and how close the fire is to the road. Helicopters provide a versatile transportation option that can deposit a crew close to a fire site (if a suitable clearing is nearby), but they are expensive when compared with trucks or fixed-wing aircraft and have limited load capabilities (depending on the size of the machine – and larger machines can be very expensive). A variant on helicopter transport is the rappel crew, who rappel from a hovering helicopter, through the trees to be dropped off very close to the fire site. Rappel deployment introduces its own set of hazards for the firefighters, and specially trained and outfitted crews are essential for success, but it offers the ability to deploy crews very close to fires, despite a lack of clearings or very tall trees.

Machines: on the ground When fires get big, and the fight becomes protracted, heavy equipment is often brought in when the ground conditions support it. There are a number of options but most often used is the bulldozer (Fig. 7.8). These large pieces of construction equipment can quickly build wide fireguards in flat to gently rolling terrain when there is sufficient soil that is not too rocky. A wide range of other heavy equipment, both standard construction and specially designed, is pressed into service but the basic aim is always the same; to separate burnt from unburnt fuel, creating a fire break.

Aerial-based resources

Aircraft are used extensively to directly combat forest fires. Their ability to travel rapidly and move water quickly to any point on a fire perimeter makes them an invaluable resource to the fire suppression team. Further, the large capacity of some

Figure 7.8
A bulldozer fire line established in the Canadian
Rocky Mountains. Note the connection of the fire
line to the road in the foreground providing a
continuous break in fuels. Photo courtesy of John
Niddrie, Parks Canada.

air tankers can effectively cool a high-intensity fire down enough so that firefighters
can directly attack the fire front. There are many considerations in the selection
of an air tanker fleet; airport network density, water availability, and aircraft maintenance and reliability.

The selection of specific air tankers depends on the details of a mission profile.
A mission profile lays out a range of expected situations (or specific capabilities)
that the fire-management agency wishes an aircraft to be able to respond to effectively. Mission profiles might include:

- An operating radius

- Minimum response time to a site a specified distance away

- Minimum residence ('loiter') time at the fire site (see below)
- Minimum drop capacity and configuration (see below)
- Multiple drop characteristics
- Manoeuvrability in tight terrain (some mountainous terrain can restrict aircraft selection)

Most elements of the list above are fairly straightforward; however some further details on a couple of items are warranted. In the case of residence time or loiter time, this refers to aircraft that stay around the fire site and continue to pick up and drop water from local sources (Fig. 7.9). Residence time is governed by a number of factors, but most notably aircraft fuel capacity when bombing. Typically the more fuel on board the aircraft, the less water it can carry since aircraft have a maximum weight-carrying capacity.

The second element bearing closer scrutiny is the drop configuration. When air tankers release their loads, a multitude of factors interplay, but the result is water on the ground. The pattern of that water is called the drop footprint, and is depicted for analysis as a set of lines similar to contour lines on a map. These lines describe lines of equal 'depth' of water deposition. It is easy to imagine the elements contributing to the footprint pattern: aircraft height and speed at the time of load release, wind speed, and the configuration of the tanks and tank doors on the aircraft (how fast and lumped together does the water come out?). All of that aside, it is the footprint produced that interests the fire manager (optimising the footprint for any given aircraft is the business of the aircraft engineer). The fire conditions at the time of the drop will dictate what depth of water or retardant (see Box 7.4) is required to suppress the fire. Too little depth and perhaps the load may not penetrate the forest canopy to reach the fire below, or even if it does, it does not reduce the fire intensity enough to be effective. Too much water depth is a bit of a waste – the drop could have been stretched out further (at a reduced but still effective depth) to increase the fire-line building production rate; or worse yet too big a load in a wildland – urban interface fire can cause damage to homes (a whole new definition of water damage).

Changing the timing of the tank doors opening[2] and changing the drop speed can change the drop footprint to produce a result appropriate to the fire conditions. This, of course, has to be within the capacity of the given aircraft; nothing can make an air tanker effective on a high-intensity fire when the load is just too small. However, there are other alternatives; we have seen in Chapter 3 that fire intensity varies around the perimeter – highest at the head and lowest at the

Figure 7.9
Bombardier CL-415 water bomber aircraft dropping a practice load. A load of water can be scooped from a lake in less than 10 seconds while the aircraft taxis across the waterway. Ontario Ministry of Natural Resources, copyright 2009 Queens Printer Ontario.

---

2. On larger air tankers the water or retardant is stored in a number of onboard tank compartments, each released by its own door. Thus the timing of the door openings (for example, all at once or a one-second delay) affects the drop footprint.

back. Often when an attack on the head is fruitless, air tankers can be effective on the flanks, where their load can limit lateral spread in strategic locations.

A plethora of options exists for aerial water delivery to fires. The options are not mutually exclusive – often one will see a mixture of aircraft types and models on any large fire. The first big split is between the helicopter group (or rotary wing) (e.g. Fig. 7.10) and the airplane (or fixed wing), each with its own set of positives and negatives, capabilities and shortcomings. The second major split in fixed-wing aircraft is the land-based retardant aircraft and the skimmer air tanker. Land-based retardant aircraft load a water-based mixture of salts and clay at an airport, and fly to the fire (or fires) to drop the load, while skimmer aircraft pick up water from lakes and drop it on the fire. Each type has a number of variations, models, capacity and capabilities. Box 7.3 outlines some of the strengths of each type.

Step 3: suppression

We have seen above that there is a tremendously wide range of resources that can be sent to a fire, and this is where the fire-management job becomes both very simple and very complicated.

First of all it is simple: to extinguish a fire one needs only to break the fire triangle described in Chapter 3. Remembering that this fire triangle is composed of fuel, oxygen and heat – a firefighter needs only to remove one of these to suppress the fire. Remove the fuel from in front of the spreading fire, and there will be nothing to burn. Cool the fire sufficiently (with water, foam or gel – Box 7.4) and there will not be enough energy to preheat unburnt fuel to the combustion point and nothing further can burn. Starve the fire of oxygen by burying or submerging and it will be unable to sustain combustion. The theory is simple and in many cases, when the fire is small and low intensity, putting the theory into action is also simple; arrive at the fire, build a line around the fire to stop the forward spread, and then extinguish the fire. The 'line' around the edge of the fire is the fire line – a line where further fire spread has been prevented, by breaking the fire triangle. The most common type of fire line is a physical separation or break in the fuel complex (imaginatively named a fuel break) whereby unburnt fuel is broken and scraped away so that a line, absent of fuel, is created (see Box 7.5 for more details). Once a line is established around a fire it is strengthened and patrolled so that the fire is not able to jump across the line, and the interior of the fire is then extinguished.

Unfortunately, while at its essence the concept of fire suppression is simple, things can get complicated very quickly. Over 90% of all wildland fires are successfully suppressed while they are small, and the suppression procedure is straightforward. But when high-intensity fires erupt, suppression techniques and tactics change to

Box 7.3. The strengths and weaknesses of different aircraft used in fighting forest fires

Aircraft category	Class	Operations	Pros	Cons
Helicopter Fig. 7.10	Bucket	Helicopters with buckets hanging below dip into available water sources and deliver the water to locations on the fire line. In some cases foam concentrate (see Box 7.4) can be injected into the water in the bucket to increase effectiveness.	Pick-up can be done in very small natural or artificial water sources. Water can be delivered to precise locations by helicopters hovering over the target and releasing the load, or it can be spread over a larger area depending on how fast the helicopter is travelling. Helicopters can also be quickly used for a variety of other purposes – moving personnel and equipment around – making them highly versatile.	Helicopters are slower than fixed-wing aircraft, so if the water source is fairly distant from the fire, ferry time can be considerable. Capacity is limited in most aircraft – clearly the larger the helicopter, the larger the load capacity, however large helicopters are expensive to operate and are relatively scarce. Common 'medium-sized' helicopters (e.g. Bell 205) can carry 1350 litres while a light helicopter (e.g. Bell 206 long ranger) can carry 410 litres. Buckets must be removed before an internal load is taken (for example, people or equipment in the cabin).
Helicopter Fig. 7.11	Other	Recent adaptations to firefighting helicopters include 'belly tanks' which turn helicopters into a closer cousin to the air tanker.	Helicopters can travel faster without a bucket hanging below the aircraft, but speed is still limited compared with conventional fixed-wing aircraft. Helicopters with belly tanks attached can also carry internal loads.	See above.
Fixed-wing Fig. 7.12	Land based	Land-based air tankers load up at an airport, fly to the fire (or fires) and deposit their load. The aircraft are typically converted commercial or military aircraft that have been given a second life through retrofitting.	The aircraft can be very large with a tremendous capacity, and often can split their loads up, depositing parts on several targets. Typically the aircraft carries fire retardant, a slurry of water, clay and salts that is much more effective than straight water at suppressing a fire. Smaller air tankers, similar to agricultural crop dusters, are also used when a reloading location can be established close by the fire. The sheer range of possibilities of air tanker size for this class is an advantage, allowing fire-management organisations the opportunity to choose the capacity and characteristics of the aircraft to match their needs.	Land-based air tankers are the standard used across the USA and western Canada. The major drawback for the technology is the long turn-around time for the aircraft to reload after making a drop. The tanker must fly to the closest airport with reloading facilities. This then requires an extensive ground support programme and also intensifies the importance of hitting the target with a drop – a miss can mean a wasted trip and an escaped fire.
Fixed-wing Fig. 7.13	Skimmer	Skimmer air tankers are water-based or land/water-based aircraft that can skim a water body to fill onboard tanks, fly to the fire, drop the load and return to a nearby lake to refill. Originally these aircraft were	Skimmer aircraft can be extremely effective at depositing tremendous volumes of water (or foam-enhanced water) on fires in short order – if there are lakes or sea nearby. Newer tank	Ex-military aircraft are World War II vintage and have extremely limited lifespan left, if indeed any are still flying. Float-based skimmer aircraft typically have limited water capacity and may not be

Box 7.3. (cont.)

Aircraft category	Class	Operations	Pros	Cons
		ex-military, adapted for a new role. The PBY-5A submarine hunter is the classic example of an adapted water bombing aircraft. Other aircraft adapted to the skimmer role are float-based fixed-wing aircraft that use their floats as water tanks to pick up and drop water. Recently (since 1970) a specialised skimmer water-bombing aircraft was designed and built by Canadair. Their CL-415 is the only commercially produced aircraft designed and built for fire suppression.	design allows computer-controlled tank door opening, thus controlling the 'footprint' of the drop on the ground.	suitable for fire situations where the load must penetrate a forest canopy, or in high fire-intensity situations. The Canadair models (CL-215, and CL-415), while extremely capable in areas dominated with lakes, are not as effective in more arid landscapes. They are somewhat slower and have a limited water capacity when compared with their land-based cousins (the larger ones at least).
Fixed-wing Fig. 7.14	**Other slip-on tanks**	A unique category for air tankers is the convertible cargo plane. The MAFFS (Mobile Aerial Fire Fighting System) slips aboard a military cargo aircraft converting it, in a matter of minutes, from a cargo plane to an air tanker. It becomes similar in many respects to a land-based air tanker once loaded.	Most firefighting aircraft are very specialised, and as such are used for little else during the off season. Slip-on tanks allow aircraft to perform multiple roles throughout the year.	Limited numbers of aircraft are available in many countries for this kind of rapid retrofit. Other constraints are the same as other land-based air tankers.

Figure 7.10
A US Navy MH-60S Seahawk helicopter drops water from a 420-gallon bucket slung underneath the aircraft onto one of the many areas affected by wildfires in San Diego, California, 23 October 2007. The bucket can be dipped into a water source (lake, pond, reservoir, swimming pool etc.) and quickly transported to the fire site. US Navy photograph by Mass Communication Specialist 2nd Class Chris Fahey.

Box 7.3. (*cont.*)

Figure 7.11
Kern County Fire Department (California, USA) Bell 205 helicopter dropping water from a belly tank during a training exercise. Most belly tank helicopters are equipped with a 'snorkel' that when deployed hangs down under the aircraft during flight and sucks up water to fill the belly tank. Photograph posted to Wikipedia Commons, courtesy Alan Radecki.

Figure 7.12
A heavy air tanker contracted to the US Forest Service delivering 3000 gallons of fire retardant. Photo courtesy of Chris Carlson, North Carolina Forest Service and US Fish & Wildlife Service.

Figure 7.13
Bombardier CL-415 water bomber aircraft dropping a load. Photo courtesy of Mitch Miller, Ontario Ministry of Natural Resources, copyright 2009 Queens Printer Ontario.

Box 7.3. (*cont.*)

Figure 7.14
C-130 Hercules aircraft equipped with modular airborne firefighting system (MAFFS). The aircraft can drop up to 14 000 litres (3000 gallons) of retardant covering an area 0.4 km (one-quarter of a mile) long and 18.5 m (60 feet) wide. Courtesy of the US Air Force.

Box 7.4. Water, foam retardant and gel

Water, through the millennia, has been the preferred method for cooling and extinguishing a fire. During the mid 1980s Class A foams were developed for wildland firefighting. Foams are essentially surfactants that lower the surface tension of water and allow it to better cool and penetrate fuels, while the physical layer of foam also limits the oxygen supply to the combustion zone: where water would hit and run off, foam-enhanced water will expand and cover the fuels. Foam concentrate is added to water in extremely low concentrations (typically less than 1%) and are considered environmentally benign and safe for the environment. Fire retardant is a longer-term fire suppressant typically delivered by land-based air tankers. Retardants are composed of a phosphate, clay thickener and colourant. The phosphate helps the other ingredients mix with the water by reducing the surface tension of the water (allowing it to be better absorbed by the ingredients). The phosphate also acts as a soil fertiliser aiding plant growth after the fire. The clay holds the moisture of the water for an extended period, releasing it slowly and so reducing evaporation losses and extending the effectiveness. The colourant makes the drop visible from the ground and the air, allowing successive air tanker drops to build a continuous fire line. Retardants are considered long term in that their effectiveness will last many hours (as opposed to water or foam). Firefighting gel is a new development from the early 2000s, and is targeted primarily at protecting structures in the wildland environment. The chemical ingredients, when mixed with water, form a persistent gel that will adhere to vertical walls and absorb a great deal of heat energy without transmitting it to the structure. Gels have not yet been accepted into regular use by fire-management organisations, but they show great promise for the future.

Box 7.5. Fire lines – fuel breaks

Fire lines or fuel breaks differ greatly for the vast variety of conditions of soil, fuel and fire intensity in which they are used, not to mention the equipment used to construct them. They can be tool lines that are centimetres wide, or they can be constructed by multiple bulldozers and be tens of metres wide. There are many considerations to their design – first the fire: how intense is it? This will define (to a certain extent) how wide the line is. Intense fires have high levels of radiative heat and the ability to produce profuse short-range spot fires, thus requiring wide fuel breaks. Lower-intensity fires can be contained with narrower.

Soil or ground type can clearly impact the ability to build fire lines. Flat sandy soils are easiest, steep rocky terrain is very difficult and can preclude the use of machinery, as can swampy areas. Peat lands require special considerations – here the fire can burn deep underground requiring excavation to ensure it is extinguished.

Fuels also impact the fuel break. The physical nature of grass fuels or low shrub fuels can (all other things being equal) require narrower fuel breaks because of a lack of overhanging fuels found in forests. Trees burning inside the fire line can fall and breach a narrow fuel break unless precautions are taken. Fuels with a loose hanging bark or other spot fire material can require wider fuel breaks and require more frequent patrolling of the fire line to ensure no jumps have occurred.

meet the challenge. There are two basic tactics for establishing the fire line: direct and indirect attack. A third tactic (parallel attack) is a cross between the two.

Direct attack is when firefighters work directly on the flaming fire front. Indirect attack is the construction of a fire line some distance away from the flaming fire front and (most commonly) burning out the intervening fuel between the constructed fire line and the flaming front. Parallel attack (a subtype of the indirect attack) is the construction of a fire line close to, but not at, the flaming fire front and either burning out the intervening fuel or allowing the fire front to advance to the constructed line. Each of these methods has a place and all can be used on a single fire, remembering that the perimeter of a fire can exhibit a wide variety of intensities, depending on the fuel and weather conditions and the position along the perimeter (head, back or flank of the fire).

Direct attack is used on lower-intensity fire fronts where the heat and resulting firefighter stress is tolerable. Ideally firefighters arrive soon after the fire has started, while it is still small and low intensity. Here the crew can directly attack the fire front, and quickly contain the fire before the intensity becomes too high. If possible the crew start at the head of the fire where the fire is most intense and it is spreading fastest. By attacking here, growth of the fire is arrested most quickly, the total fire area is maintained at a minimum, the fire front perimeter is minimised and finally the length of fire line to be built is also minimised. The actual upper limit of fire intensity endurable by front-line fire crews in direct attack varies (by fuel complex, rate of spread and fire-line equipment) but is generally thought to be less than 2500 kW/m (see Chapter 3 for an explanation of fire intensity). Air support can extend this upper range by several thousand kW/m by dropping suppressant (water, foam, retardant) directly on the fire front, thereby cooling the fire, reducing the intensity and providing an opportunity for direct attack. Hirsch *et al.* (1998) documented the impact of having a bucketing aircraft assisting ground crews and was able to measure the improvement in the probability of containment during initial attack. The fire-intensity reduction from air attack can be short-lived depending on the original fire intensity and the characteristics of the drop. Extremely high-intensity fires can quickly burn through retardant, or evaporate dropped water to return to their original intensities. Additionally, ground crews working in concert with air support have to work both at a safe distance (to avoid injury from the drop) and close enough that the efficacy of the drop is maximised (by getting back to the fire line while the drop is still influencing fire behaviour). Large air-tanker drops can easily snap trees and disperse debris when the load hits, making the drop zone a dangerous place to be.

Indirect attack is used in a number of situations to mitigate firefighter safety concerns, reduce cost and manage the overall suppression effort. When the intensity of the fire is too high (greater than roughly 2500–3000 kW/m) the heat and stress of working directly at the fire line can be too great for firefighters, and line construction is slowed dramatically. At this point, fire crews will switch to an indirect attack. Indirect attack is typically used on fires that are larger than a couple of hectares, are burning aggressively in lower-value fuels (i.e. not houses) and would consume a great deal of resources to attack directly (if this was possible). Essentially firefighters work a distance away from the main fire front, ideally taking advantage of existing topographic features such as fuel breaks and then either waiting for the fire to reach this fuel break, or by burning the intervening fuel between the fire break and the fire front, effectively widening the fire break and starving the fire of fuel. The distance separating the fuel break from the existing fire front varies tremendously, from a few tens of metres (sometimes called parallel attack) to kilometres for very large fires, depending on the strategic and tactical situation. Ground crews may never actually be involved in the process; the fuel break used may be a stream, road or ridge top, and the intervening fuel burnt off by aerial ignition (described later under Step 4) in the late afternoon or evening (when burning conditions have lessened, allowing for a large area of land to be burnt off with little risk). In this way large areas of land (several hundreds of hectares) can be burnt to a stable perimeter with little risk of escape. When indirect attack is utilised without burning out[3] and the fire front allowed to approach, ground crews have to be vigilant to prevent the fire from burning over the fire line, spot fires from jumping the line, or debris (trees or logs) from falling or

---

3. Burning out, back burn, back fire and burn off are all largely synonymous terms used to describe the act of igniting and burning fuel situated between a control line and the active fire front.

rolling (down a slope) across the line. In fact, vigilance is always required during and after indirect attack operations to prevent the fire line from being breached.

Combined methods of direct and indirect attack are commonly found during suppression. Although fire growth is commonly represented by an ellipse, fires rarely produce a smooth fire line (see Chapter 4). Typically the fire edge is a mixture of fingers and bays twisting around the theoretical smooth boundary. Much twisting can be related to fuel and topographic issues, but has been found to be a fractal dimension (see McAlpine & Wotton 1993) – the degree of 'twistiness' is similar when measured at a variety of scales. Employing a direct attack on the entire twisting edge would be challenging, and somewhat wasteful; rather fire crews, during the initial containment stage, will directly attack the forward edge of the aggressively burning head, and cut across unburnt bays[4] (depending on the size of the bay) to quickly establish a control line completely around the fire. These internal bays can then either be left for the fire to burn out, intentionally burnt off, or in subsequent suppression passes extinguished through direct suppression (Fig. 7.15).

Aircraft add another dimension to fire suppression and their tactical uses are as varied as the fleets and fire situations faced. While firefighting aircraft can extend the range of intensities at which fire crews can directly attack fires, they perform a number of other tasks to suppress fires. Typically, when working a fire edge for direct suppression, air tankers will seek to drop one-third of the load on the fire and two-thirds of the load in front of the fire edge. This immediately cools the fire and retards further growth (Fig. 7.16). As stated above, air tankers can first cool a fire front allowing a crew to work directly on the edge of the fire and effectively suppress it. This is most common during the early stages of an initial attack to allow the crew to establish a line around a small but aggressive nascent fire. When a fire is difficult to get to (due to remoteness or rugged terrain), fire crews may take some time to actually arrive at the fire, in which case air tankers can slow the fire and contain it until the crew arrives. In this case, a load (or multiple loads depending on the air tanker) is dropped on the fire, and the tanker returns to base while the crew travels to the fire. Indeed this can be so effective that the crew can have great difficulty in finding the fire when they arrive. Air tankers can build a fire line for indirect attack when using retardant. A series of drops in unburnt fuel establishes a continuous line of fire retardant on the ground from which a burnout can be performed. This technique is more advanced (and complex) and used on larger fires where direct attack is impossible (or unsafe) and ground fire-line construction would be too time consuming. On larger fires, aircraft can be called upon to cool sections of fire line that are flaring up, giving the ground crews a much needed hand. Because of their speed, they can move around the entire perimeter of a large

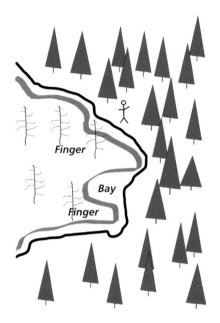

Figure 7.15
A stylised fire line (red) showing characteristic fingers and bays typical in free-growing fires and a typical first containment fire line (black) cutting across the unburnt bays to contain the faster-moving fingers.

Figure 7.16
The ellipse shows an ideal target location for an air tanker drop in this scenario placing one-third of the load on the fire front and two-thirds of the load on the unburnt fuel. Ontario Ministry of Natural Resources, copyright 2005 Queens Printer Ontario.

4. An unburnt bay is formed between two 'fingers' of faster fire spread. The material left in the bay may be unburnt for any number of reasons – perhaps a wetter area or different fuel type.

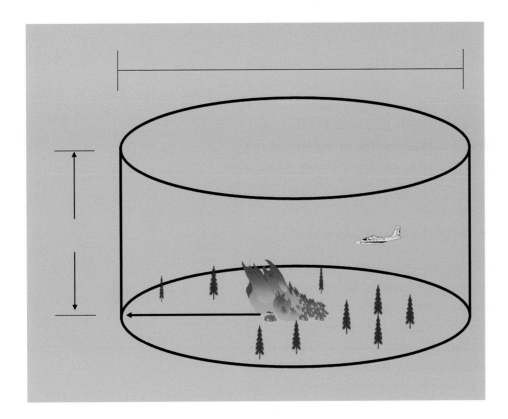

Figure 7.17
Bird dog aircraft establish an air traffic control zone around a fire to manage aircraft working around the fire. The control zone typically has two axis – a radius that extends well beyond the perimeter of the fire as well as a vertical dimension extending many hundreds of metres up from the ground surface to define a large flat cylinder. Ontario Ministry of Natural Resources, copyright 2009 Queen's Printer Ontario.

fire (which can extend to many kilometres) and deliver water where needed. This is especially true of helicopters fitted with buckets or belly tanks. Their ability to hover and deliver payload with pinpoint accuracy can help tremendously with localised problems. When water is in short supply, aircraft can ferry water from a source and drop it off in staging tanks for crews to use as required. This again is primarily the role of the bucketing helicopter as it can ensure the bulk of the load is delivered into the tank.

Air tankers are directed where to drop their loads by the fire boss or incident commander. They are in control of the suppression forces being used on the fire and establish the strategy and tactics to be used to achieve a successful attack. A second class of aircraft is often involved too. The incident commander is aided by the 'bird dog' aircraft. This smaller, lighter and faster aircraft has better visibility and is more manoeuvrable than the larger air tankers. These characteristics allow the bird dog to perform four primary activities on the fire: lead large air tankers through the approach and indicate the target for the tanker; provide feedback to the tanker on the success or accuracy of the drop; provide information about the fire's growth, extent and intensity to the incident commander when they are on the ground; and on large fires act as an air-traffic controller managing all the aircraft working on the fire to ensure safe separation between aircraft is maintained (Fig. 7.17). So while the incident commander is in charge of the suppression effort, he/she is aided by a pair of eyes in the sky who can help provide intelligence on the fire activity and direct the aerial assault.

To sum up: suppression forces arrive at a fire, typically (but certainly not always) ground crews arrive first and the incident commander scouts and assesses the fire. A suppression strategy is decided upon (direct, indirect, parallel) and the need for additional resources is considered. Air tankers (if required and available or present), led by a bird dog aircraft target aggressively burning sections of the fire front to allow fire crews to contain the fire or build a retardant fire line from which a burnout can be conducted.

Step 4: suppression failure – large fire management

What happens when despite your best efforts the fire escapes and grows larger? As we have seen there are just some days when the weather, fuel and topography all come together with the ignition source to produce a fire that cannot be stopped. On these days nature has her way. During the early stage of extremely aggressive fire growth, no amount of suppression effort will be successful at stopping the spread of the fire. The best that can be achieved is to fall back, evacuate the people in the path of the fire (or not – see Box 7.6) and attempt to protect the values from destruction.

Box 7.6. Fire evacuation?

When a fire is spreading aggressively and is beyond the control efforts of the suppression organisation, it would seem obvious to evacuate the people from the path of the fire and stand back. Most organisations follow this protocol; however in Australia this has not been the standard practice. Rural homeowners in Australia are encouraged to have a plan to 'Prepare, Stay and Defend, or Leave Early'. Research results from Australia indicate that most fatalities are not the result of residents being trapped in their homes, rather it happens when they are fleeing the conflagration. A wildland fire will pass over a home in a few minutes, and if the home is properly designed and prepared, it will survive and protect its inhabitants. However, residents fleeing can easily become disoriented and have no way of knowing where a safe location. The key to the Australian doctrine is to plan ahead and stick to the plan. Individuals who plan to evacuate and do so early do not become disoriented in the smoke and chaos of the wildland fire. Individuals who stay and defend and stick to it will emerge from their homes and be able to suppress the remnant fires following the passage of the main front. The keys to success in the stay and defend strategy are a home and yard that help to inhibit fire combustion (see Box 6.3), and sticking to the plan through the fire front. Sticking to the plan is perhaps the most difficult aspect, as the passage of a fire front over a home is an extremely frightening event. Recent events (2009) in Australia have prompted a review of this policy and the challenges it presents. This review may be a result of society becoming more urban (and less rural), and wildfire becoming more of a threat to urban areas.

Value protection takes many forms, again depending on the value being protected and the suppression resources available. Obvious steps are to wet down the structure with water, foam, retardant or gel-based fire inhibitors. In particularly valuable cases (such as historic buildings), and where there is enough time, the structure can also be clad in foils or aluminised boards which will help to reflect radiative heat. If there is sufficient time, fuels from around the structure or facility can also be cleared to reduce the intensity of the fire immediately around the location of interest. Canadian fire-management agencies have been successful using a temporary water sprinkler network around structures. These rapidly deployed sprinklers supplied by a standard wildland fire portable pump are highly effective at preventing structures from burning during the passage of the fire front.

Large fire suppression follows the same basic practices as those used on smaller early-stage fires. The difference is in the scope and scale of the operation. Instead of from perhaps 3–30 people fighting the fire, there could be hundreds or even thousands. When fires get that big, large organisations are required to coordinate, supply, feed and pay the firefighters. To control so many people, the fire perimeter is divided into divisions and sectors and are devolved to division supervisors, sector bosses and on down to crews. The organisational structure very closely parallels a military operation, and beyond the front-line firefighters and aircraft there are sections within the fire command structure to manage plans and records, and sort logistics, finance and administration. The fire organisation takes on the function and structure of a fire headquarters with a myriad of personnel supporting the operation.

Burnouts on large fires can be contemplated on a much grander scale than small fires. Strategic burn blocks planned to contain a large fire can measure hundreds of hectares in size, and be ignited with aerially mounted ignition devices. The two most common aerial ignition devices are the heli-torch, which drops ignited jellied gasoline (Fig. 7.18), and the DAID (Delayed Aerial Ignition Device) machine which drops balls the size of ping-pong balls partially filled with potassium permanganate and injected with ethylene glycol (antifreeze) or glycerine just prior to release (the potassium permanganate and ethylene glycol produce a flaming combustion reaction after about a 20-second delay). Either aerial ignition system can spread fire over a large area quickly, but each is suited to different fuel conditions and desired results. Most often these large ignition projects are conducted in the early evening to limit the time that the burnout can aggressively burn before the cool moist night air limits the fire intensity. The advantage of the aerially mounted ignition device is that large areas can be burnt off in a relatively short period of time with little risk to firefighters on the ground.

Figure 7.18
A helicopter uses an aerial drip torch to drop gelled gasoline during a burnout operation. Photo courtesy of Mitch Miller, Ontario Ministry of Natural Resources, copyright 2007 Queens Printer Ontario.

Fatality fires

It is an unfortunate fact that sometimes when men and women fight fires there are fatalities. Often these fatalities are the product of accidents that could happen in any similar industrial situation – vehicle accidents, aircraft accidents and heart attacks are the most common. However, sometimes firefighters lose their lives to the fire. Many studies have been conducted following these fires to look at the reasons why these tragedies occur, and a review by Wilson (1977) reveals the bulk of the commonalities. It may seem odd but the high-intensity, fast-moving fires we see on the news are not the fires that cause firefighter fatalities. These fires are an obvious threat and the power and potential of the fire is well respected. Wilson found the common elements of fatality fires to be:

- Small fires
- Light fuels
- Wind shifts
- Up-slope fires

These common elements all link to one common issue – firefighters underestimating the fire potential.

Sometimes small fires, early on in their development, have not yet reached equilibrium spread with the weather and topographic conditions. Perhaps the fire is burning in a localised wet area, or a small hollow where it is protected from the wind. When it hits some trigger that quickly transforms it from a small relatively benign fire to a rapidly moving aggressive fire, firefighters can be caught off guard. Fires burning in light fuels are often underestimated for the simple reason that they are light fuels – they seem innocuous enough. But light fuels (for example grass) have the greatest potential for extremely rapid spread and react quickly to changes in wind speed and direction. This rapid reaction to wind changes can take firefighters by surprise.

Clearly wind shifts or changes in direction, unless planned for or anticipated, can quickly change a situation and threaten firefighters. Slope has a similar effect, although for a couple of different reasons; fires moving up a slope travel much faster than on flat ground so firefighters above a fire can quickly get into trouble. Here a second confounding factor should be pointed out – fires can spot behind the firefighters to perhaps a lower part of the slope and catch the crew in a poor position. This was the case of the 1949 Mann Gulch Fire in the western USA where 13 smokejumpers died racing in front of a fire burning in light fuels up a steep slope (Rothermel 1993).

Some agencies distribute 'fire shelters' to firefighters. These are small, one-person, lightweight tents essentially made from a type of aluminium foil. The fire shelter is the firefighters last best chance for survival when overrun by fire. Fire shelters are effective in some scenarios – when they can be deployed on bare ground (no fuel) and when they are deployed in a clearing large enough that the radiant heat can be managed by the shelter. Much is done to prevent firefighter fatalities, including improved training, improved equipment and improved understanding of fire behaviour but the work will always be dangerous and fatalities may be inevitable.

The fire-management organisation

Planning ahead

We began this chapter with a discussion of the fire intelligence function and how it sets the stage for fire detection, dispatch and suppression efforts. In fact the fire intelligence function goes beyond the here-and-now of daily planning and reaction to fire ignitions; it extends to planning for activities beyond today into tomorrow and further into the future. Fire-management organisations do not only react to the situation set before them, they take an active role in assessing the fire load to be expected in the coming days, both from ongoing fires and expected new starts. By reviewing the expected fire behaviour and fire occurrences over the region, the fire-management organisation can move resources around to meet anticipated need, pre-positioning them to reduce travel time to new fires when they start and thereby maximising effectiveness and ultimately suppression success. This process can also identify when the organisation is falling short of adequate resources and should consider requesting additional resources from a neighbouring or partner agency. Conversely, the assessment process allows the organisation to consider requests from other agencies for suppression resources.

To do all this, relatively static information (like fuel-type maps and topography) is spatially overlaid with weather forecasts and processed through fire danger, behaviour and occurrence models to produce a map of fire-load expectations for several days into the future. Clearly the weakest part of the forecasting function is the weather forecast, as the accuracy plummets quickly when looking beyond 2–3 days. The plan formulated based on this information, however, is just that – a plan – and it is modified and extended each day. Understanding the limitations of the forecast and the inherent uncertainties is part of the planning cycle. How far into the future the fire-management team works depends on the spatial extent and responsibilities of the group. Local or district-level planning may be limited to 24+ hours, region-wide planning (integrating the plans of many districts) will extend at least to 48 hours, while agency-based plans (encompassing multiple regions) may look forward 3–5 days or beyond.

Planning beyond the here-and-now

Prior to the 1970s wildland fire organisations were called fire-control organisations. During the 1970s, however, it became increasingly clear and accepted that fire was a natural part of the ecosystem process, and fire-control organisations renamed themselves fire-management organisations. With this change came some simple reconciliations and realisations. Where fire-control organisations operated like a city firefighting service, extinguishing all fires, fire-management organisations seek to balance the economics of human needs with the ecological needs of the forest

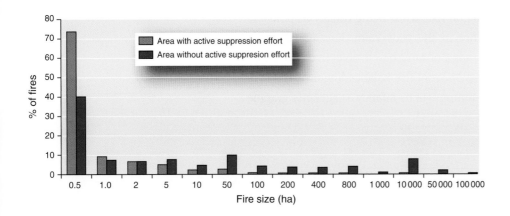

Figure 7.19
Fire size distribution of an area with aggressive full suppression policy, and an area with a near natural fire regime. While in both cases the majority of fires are small, in the natural fire area there are many more fires in the larger size categories.

and wildlands. Usually the balance is articulated in land-management objectives and plans that describe the goals for a particular area. Almost all fire-management agencies in Canada, for example, have now divided the landscape into zones with differential fire response objectives according to values at risk and land-management objectives within each zone. This practice results in some areas having a near 'natural' fire cycle where numerous fires in remote, sparsely populated areas are observed only and allowed to run their course rather than be actively suppressed. A consequence of this is reflected in the fire-size distribution of actioned vs. non-actioned fires in parts of Ontario. Fire sizes in the non-actioned zone of Ontario are much larger than in the actively suppressed zone (Fig. 7.19). This practice of letting fires burn is most evident where the land-management objectives are less linked to economic goals and more associated to ecological objectives. Parks Canada, the agency responsible for the preservation of the ecological integrity of national park land, is a prime example of an agency looking to keep fire as a natural process on the land base. By 1980 it was being pointed out to Parks Canada that either fire needed to be used as a management tool or an artificial means of vegetation renewal had to be permitted, or drastic changes in park vegetation and wildlife with time must be accepted. The first option was chosen. With a new directive produced in 1986 and a comprehensive fire-policy review, Parks Canada embarked on a new relationship with fire and continues to work to restore it to its natural role by active management. Unplanned wildfire is unacceptable in most parks because of the implied lack of control, and risks to public safety, property, rare species of flora and fauna and delicate habitats. Plus, the years of suppression had resulted in fuel accumulations and changes in fuel complexes that would have resulted in fire behaviour completely different from pre-suppression conditions. The only acceptable option was planned-ignition prescribed fires. Parks Canada has, in many parks where fire is a natural part of the ecosystem, established prescribed fires to emulate the natural role of fires while mitigating risks to ecosystems, values and public safety (Chapter 8). This embodies one of the major challenges facing fire-management organisations today: meeting the protection demands of society for values at risk, while being fiscally responsible and recognising the ecological role of fire in the ecosystem. Fire-management organisations responsible for a broader range of land-based objectives are more challenged to meet expectations of protection from fire balanced against the ecological need for fire. Indeed this level of planning is in its infancy as agencies grapple with competing priorities and viewpoints.

8 | Wildland fire and its management – a look towards the future

Kelvin Hirsch

The future ain't what it used to be – *Yogi Berra*

The age of uncertainty

In 1899 Gifford Pinchot, who in 1905 became the first Chief of the United States Forest Service (USFS), wrote an article for *National Geographic* magazine entitled 'The relation of forests and forest fires' in which he stated, 'That fires do vast harm we know already, although just what the destruction of its forests will cost the nation is still unknown.' Influenced by European views of forestry and founded on the belief of human mastery over nature, the passionate and charismatic Pinchot, along with his understudies and eventual successors, led a crusade against fire that was considered both morally acceptable and economically desirable. Forests, like other natural resources, were deemed valuable assets to be used to meet immediate human needs and increase the wealth of those who owned them. Pinchot's conservationist vision 'to control the use of the earth and all that therein is' (Cortner & Moote 1999) was confidently thought to be not only plausible but completely possible owing to an era of tremendous optimism in western society at that time. Reaping the benefits of the industrial revolution, including the steam engine, electricity and the telephone, human progress was fuelled by a philosophy of scientific determinism that viewed the world and how it functioned as orderly, machine-like, and therefore predictable as well as controllable. The famous author H.G. Wells captured this sentiment in 1902 in his address at the Royal Institution of Great Britain suggesting that if humans were to channel the same level of scientific effort on the future as had been placed on understanding the past it would be possible to predict what lies ahead with certainty. Thus even after the Big Blowup Fires of 1910 (Pyne 2001b), which burnt over 1.2 million hectares in northern Idaho and western Montana and were pivotal in solidifying the direction of wildland fire-management policy in the USA and influencing it worldwide, Pinchot's (1910) conviction that 'forest fires are wholly within the control of men' faced little more than a smattering of opposition. Most forestry professionals, policy makers and even the general public believed it was not a matter of if fire could be controlled but simply when. This ideal, reinforced by the success of fire suppression in some ecosystems, was tightly held well into the 1970s (see Brown & Davis 1973).

Fast-forward to the dawn of the twenty-first century and a new millennium. A great deal has happened in the past 100 years – global population has quadrupled, reaching close to 7 billion people; we have not only conceived how to fly but have gone to the moon and are sending robotic devices to explore the cosmos beyond; and at the speed of light we can communicate with people in any part of the world. We continue to see unprecedented technological growth that has significantly improved the quality of life for many but simultaneously has also given humanity the potential for self-elimination for the first time in history. We are aware that eighteenth and nineteenth century Newtonian science, or positivism, is essential but not sufficient for making sense of our complex world and thus have commenced exploring new sciences such as chaos theory, quantum physics and self-organising systems (Wheatley 1999) in an attempt to better understand our surroundings and how we can effectively function in them at a time of accelerating change. There is widespread awareness of the importance of our environment to our individual health and well-being; natural resource management philosophy now speaks of sustainable development; and in wildland fire management there is growing recognition that it is neither economically possible nor ecologically desirable (as discussed in Chapter 5) to eliminate all fire from the landscape (Stocks & Simard 1993).

As this chapter is being written the optimism that was so vibrant just a century ago has greatly eroded and continues to falter due to wars, terrorist attacks, infectious disease outbreaks, international economic crises, corruption, human rights violations and natural disasters such as hurricanes, floods and earthquakes. Pollution and other forms of environmental degradation have transformed the dreams of a prosperous tomorrow into prophecies of anarchy and societal collapse (e.g. Diamond 2005, Dyer 2008). Adding to our despondency, rapidly increasing complexity arising from a myriad of factors including population growth, competition, interconnectedness, feedback loops between natural and human systems, and multiple dimensions is generating confusion and surprises (Homer-Dixon 2001, Lui *et al.* 2007). Uncertainty and the escalating awareness of our inability to predict the future – be it tomorrow's stock prices, next month's rainfall or next year's government – is increasing our sense of vulnerability, powerlessness and fear. Nevertheless, there are those who are dedicated to the art and science of futuring and emphasise that although we may not be able to predict the future with any degree of accuracy (see Box 8.1), it is paramount that we learn from the past in order to take actions in the present that will improve our futures (Cornish 2004, McAllum 2008).

Thus, in this chapter appropriate heed will be given to the mutual fund adage that 'past performance may not be indicative of future returns' and no attempts will be made to predict the future. Instead it will begin by exploring some major technological, social, economic and environmental trends in our world and examine how those trends are influencing wildland fire and its management. This will be followed

Box 8.1. Perils of prediction

Here are a few examples of why it is important to be cautious about predictions.

- *The phonograph is of no commercial value.* Inventor Thomas Edison, 1880.
- *Heavier than air flying machines are impossible.* Physicist Lord Kelvin, 1895.
- *1930 will be a splendid employment year.* US Department of Labor, 1929.
- *The Japanese don't make anything the people in the US would want.* US Secretary of State John Foster Dulles, 1954.
- *There is no reason for any individual to have a computer in their home.* Ken Olsen, President of DEC, 1977.
- *You will never make any money out of children's books.* Barry Cunningham (Bloomsbury Books editor) speaking to J.K. Rowling (author of the highly lucrative Harry Potter book series), 1996.
- *The price of crude oil could soar to $200 a barrel in as little as six months.* Argun Murti, Goldman Sachs energy strategist, May 2008 (note that between July and December 2008 the price dropped from its peak of nearly $150 to under $35 per barrel).

by a discussion of the concept of adaptation and its merits as an approach to effectively dealing with an uncertain future. Then three examples of innovation in wildland fire management that demonstrate our capacity to overcome difficult challenges are presented. The chapter closes by describing one idealised image of wildland fire management in the future and identifies some key ingredients that if carefully sown and cultivated could facilitate its eventual realisation.

Trends and supertrends

Analysis of trends, or the long-term general tendency or direction over time or space of factors that are influencing what is happening in our world, provides a way to systematically assess the movement and inertia in our present systems in order to obtain a glimpse of where we seem to be going in the short term. Trends exist in many processes and at many levels or scales and are often interconnected. Looking at trends is similar to a meteorologist looking at global, regional and local weather patterns in order to understand what is occurring and gain insight into what may transpire. In this vein, this section will examine a number of supertrends, or powerful instruments of change in our world (Cornish 2004) and how they are influencing wildland fire and its management.

Technological progress

Technological progress has been occurring throughout human history and has helped us advance from tribes of hunter-gatherers to agricultural and then industrial societies. Cornish (2004) suggests technology was the superforce of the

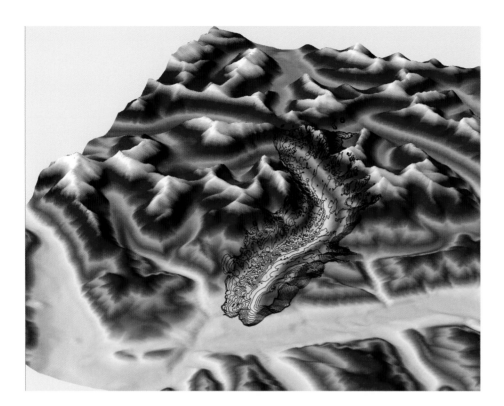

Figure 8.1
A wildfire growth simulation produced by the US Forest Service's FARSITE model. Modelling results are for the Rampage Fire which occurred in August 2003 in Glacier National Park, Montana, USA. The figure displays the simulated fire growth in one-hour time increments over a 23-hour simulation period and is draped over a three-dimensional view of topography for the general area. Courtesy of Charles McHugh, US Forest Service.

twentieth century and, barring a catastrophic event, it will continue to advance at an accelerating rate. A new technological revolution began with the creation of the computer in 1934, its continued development during and after World War II, and its eventual widespread adoption and use in our homes and offices in the 1980s. The computer has undoubtedly changed our daily lives by dramatically expanding our capability to achieve our individual and collective goals.

Likewise, computer-based technology has had a major impact on wildland fire management. In the past three decades we have garnered new abilities to inventory, assess, model and report on the fire environment. Through remote sensing we can rapidly identify forest fuels and chart topographic features (Chapter 4 explains why these are important). Automatic weather stations, remotely accessed using satellite and other forms of wireless telecommunications, allow us to measure the fire weather and fire danger conditions (see Chapter 7) on a daily, hourly and, if needed, minute by minute basis. All of this information, along with many other types of data such as values-at-risk (again see Chapter 7), is collated, rapidly analysed and presented in a readily comprehensible spatial picture using Geographic Information Systems. We have also become adept at using satellites for monitoring large-scale fire activity and smoke around the world in real-time (see Fig. 3.11) and creating high-resolution maps of fires and their impacts after they have been extinguished. Sophisticated 3D simulation models now run on computers that fit in the palm of our hand and enable us to create detailed, short-term predictions of wildfire spread and intensity (Fig. 8.1). Through Global

Positioning Systems we can track the precise location of any fire-suppression resource and dispatch it to a wildfire on a moment's notice. Computer-based technologies have given us the ability to generate and disseminate information about wildland fires like never before. The plethora of information and decision-support systems now available have enhanced the safety of those living and working in wildland areas by providing early warnings and public alerts, augmented the preparedness of our fire-suppression forces and improved the efficiency of our fire-suppression activities.

Another area where we are seeing the benefits of technological advances is related to the protection of structures in the wildland–urban interface (WUI) which is described in Chapter 6. For example, new building materials have been created that are architecturally attractive yet will not burn (for example roofing shakes that look like wood but are actually fibre–cement). Homeowners can also purchase a fire pump and sprinkler system complete with foam kits that can be triggered automatically in case a wildfire occurs while they are not at home. Or there are products that can be rapidly deployed in an emergency situation that will wrap one's house in a fire-impenetrable blanket. Steve Pyne, the well-known fire historian and commentator, in his 2004 book *Tending Fire*, indicates that technological solutions exist to our current wildland–urban interface challenges such that within a decade it will no longer be the primary defining problem of wildland fire management in the USA.

With respect to fire-suppression activities themselves, although some mechanical firefighting equipment has become bigger, faster and more powerful (for example, jet aeroplanes that function as air tankers – see Box 7.2), the basic technology currently used dates back to the early to mid twentieth century (Fig. 8.2). That is to say, hand tools (axe, shovel, Pulaski, chainsaw), petrol-powered pumps and hose, and aircraft were the primary tools used in fire suppression in the 1930s and 1940s and remain so today. Some enhancements have, however, helped to significantly increase firefighting efficiency and safety. For example, the introduction of forward-looking infrared cameras in aircraft in the mid 1990s enabled pilots and air-attack officers to see through smoke and clearly identify the fire front and other aircraft, thereby increasing the accuracy of the water/retardant drops as well as improving flight safety. Firefighting foams have made it easier and safer to defend structures in the WUI as they provide improved coverage for a longer period of time and can be applied using a 'foam and go' philosophy well before the embers begin falling. Further enhancements may be in the offing as inventors explore the potential uses of unmanned aircraft and even robots for fire suppression in extreme situations. And who knows, emerging fields like biotechnology and nanotechnology may yield advances in the next few decades that are beyond anything we can currently imagine.

Improvements to fire-suppression equipment and the development and use of computer-based fire information and decision-support systems have contributed to

steady improvement in the rate of initial attack suppression effectiveness in countries such as Australia, Canada and the USA to over 90%. That is, at present at least nine out of every ten fires that are reported are successfully controlled before they reach more than a few hectares in size. Unfortunately the small percentage that do escape initial attack are the ones that make the evening news by threatening communities, destroying property and causing loss of life. These fires also account for the vast majority of area burnt (greater than 90%), are extremely costly to action (hundreds of millions of dollars can be spent on just a single fire) and continue to draw the attention of politicians and senior public servants. Furthermore, consistent with the law of diminishing marginal returns, it appears that current fire-suppression techniques and technologies are approaching their physical (and most certainly their economic) limit of fire-suppression effectiveness (McAlpine & Hirsch 1998). This is evident in the fact that even though more and more fire-suppression resources are being deployed, the effectiveness of each additional resource is smaller and smaller; consequently, regardless of how many resources we have or how much money we spend, there will always be a small number of fires that will become large and extremely intense. Hence, although technological progress has and will continue to yield many improvements to wildland fire management, it is likely both fair and safe to say that technology has not nor will it in the foreseeable future result in Gifford Pinchot's panacea of complete control of wildland fire.

Figure 8.2
Hand line construction then and now – the technique has not really changed. The picture on the left is of the Civilian Conservation Corps in the 1930s and on the right is a US Forest Service crew on the 2008 Gun Barrell Fire in Wyoming, USA. Courtesy of (left) the US Forest Service and (right) the Gerald W. Williams Collection, Oregon State University (USFS photo #407333).

Population growth, distribution and demographics

The number of inhabitants on the earth has risen dramatically over the past millennium and especially in the last century (Fig. 8.3) and this, as well as other aspects of human population, is affecting the management of wildland fire. Population growth has been attributed in large part to technological progress that has resulted in more food, better sanitation and improved health services in many, but not yet all, parts of the world. In addition, people, especially in developed countries, are living longer and, in countries such as China and India, more are entering the consumer-oriented middle class. These factors are placing tremendous pressure on our natural resources as their extraction and use is intensified in order to meet the rapidly rising demands for capital goods. It has been conjectured that for all of the world's current population to live a 'western' lifestyle between three and four earths would be required in order to supply all of the necessary products derived from natural resources (Dyer 2008). Acknowledging that even though technological advances will continue to occur and societies may adjust there is a growing concern that our natural resources are not being managed sustainably and history has shown that a scarcity of natural resources can be linked to reductions in quality of life, political turmoil, conflict and possible societal collapse (e.g. Diamond 1995, 2005, Wright 2004).

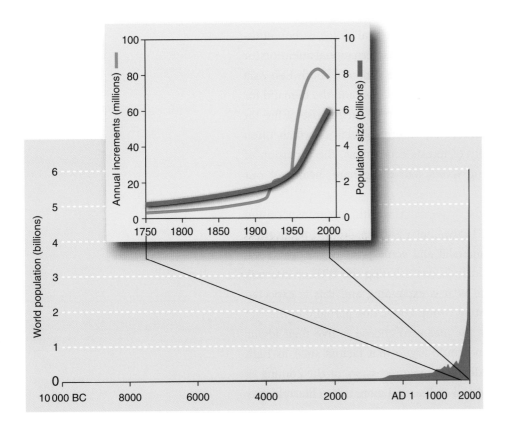

Figure 8.3
Estimated global human population since 10 000 BC. The insert shows the population explosion that has occurred since the start of the industrial revolution. Data from the UN Population Division and US Census Bureau.

Fire is a unique entity in that it can either enhance or hinder the quality and sustainability of goods and services provided by our ecosystems. In the developing world, where meeting the most basic needs for human survival is often a daily challenge for millions of people, fire remains a common tool for clearing forested land so it can be used for subsistence agriculture. It is also used to make way for commercial crops (such as coffee, sugar cane, soya and cattle) or industrial development. Its inappropriate use, particularly in non-fire dependent or fire-sensitive ecosystems such as tropical rainforests, can lead to irreversible changes in forests and other wildlands that have global conse-quences for biodiversity and climate change (Myers 2006). One specific example is from the Yucatan peninsula in Mexico where in 1988 Hurricane Gilbert levelled large areas of short-growth tropical forest. Shortly thereafter, migrant farmers and displaced families who needed food to survive arrived and burnt these sites creating enough nutrients to allow crops to grow for just a few years. Once the farmers left, the abandoned, sun-exposed sites were overrun by invasive and highly flammable grasses and shrubs and when they were accidently reburnt the sites were transformed to the point where there is now little chance of them returning to forest cover within the foreseeable future.

A second example is from Indonesia in 1997 when during a 1-year period land-conversion fires that consisted of burning peat and vegetation released an esti-mated 0.81–2.57 Gt (thousand millions of tonnes) of carbon into the atmosphere (Fig. 8.4). This was equivalent to 13–40% of the mean annual global carbon emis-sions from fossil fuel burning (Page *et al.* 2002) and drew international attention for the effect it had on human health throughout the entire region of South-East Asia (see Chapter 3). It seems inevitable that as human population and the demand for commercial products continues to grow there is likely to be further use of fire for various forms of land conversion to achieve short-term objectives which often conflict with the long-term sustainability of forests ecosystems, the provision of ecological goods and services (including water) and the general health of our biosphere.

Population growth combined with a rise in personal wealth, low transportation costs, land zoning practices and other factors have also resulted in widespread urban/suburban sprawl. In the USA, Australia and some parts of Europe much of this is occurring in fire-dominant wildlands, exacerbating the WUI or intermix fire problem. As seen in Chapter 6, the WUI is expanding and this is expected to continue (National Wildfire Coordinating Group 2009), due partially to the fact that as baby-boomers retire many have the means to choose to live their idyllic lifestyle close to nature. Conversely, one could argue that factors such as high petrol prices, 'green thinking', an extended economic recession, or the coming of age of a new generation that values time and embraces a more urban lifestyle may slow WUI sprawl.

Figure 8.4
Satellite image of haze that formed in the southern hemisphere as a result of the 1997 fires in Southeast Asia. White represents the aerosols (smoke) that remained in the vicinity of the fires. Green, yellow, and red pixels represent increasing amount of smog in the troposphere (the bottom 10 km/6 miles of the atmosphere) extending over more than 15 000 km (9300 miles). The horizontal line across the top of the plume is an artefact from joining several pictures together. Image courtesy of The Visible Earth, NASA (visibleearth. nasa.gov/).

Another interesting population-related trend is the migration of people, especially young people, from rural areas to large urban centres. As a result of this exodus, the majority of those who remain in these rural areas are elderly and are less and less able to tend the land (such as through herding livestock which helps manage fuels) or suppress wildfires when they begin. This is readily apparent in several Mediterranean countries (Vallejo 2005) and is viewed as a contributing factor to recent severe wildfire seasons that resulted in the deaths of over 45 people in Portugal and another 84 in Greece.

The last population factor raised here is the demographic shift that is occurring in developed countries and how it is affecting fire-management organisations. Data from 2005 for Canada indicated that nearly 50% of all fire-management staff would be eligible for retirement by 2015 (Born & Stocks 2006) and similar patterns exist in the USA and Australia due to the post-World War II baby boom. Considered an approaching issue for the past decade, major concerns exist about the tremendous loss of experience and knowledge within the wildland fire community as those who have proudly served their fire-management organisations for the past 25–40 years choose to contribute to society in other ways. Additionally, forest- and fire-management organisations are finding it especially difficult to entice young people, most of whom have grown up in cities, to relocate to small towns and remote regions and/or have spent a significant amount of time indoors with computers and video games to do physically demanding and potentially hazardous work outdoors in nature. In Australia this is a huge issue because most of their firefighters are volunteers and it is becoming harder and harder to find people who are willing to leave their regular jobs in the city and their families for weeks or months at a time to fight fires in rural areas and earn just enough to cover their food and accommodation (McLennan & Birch 2005). Meanwhile in the USA, the problem is further aggravated by the fact that fire-management staff are the founders and experts in the Incident Command System, a standardised, on-scene, all-hazards emergency management system (Gordon 2002), so demands on those experienced incident command staff who have not yet retired are very high (National Wildfire Coordinating Group 2009). Senior US officials are deeply concerned that if multiple natural or human-caused disasters were to occur at the same time (such as a hurricane in the southern USA and a wildfire in the west) there simply would not be enough qualified staff to address the country's emergency management needs.

Globalisation

In the early 1960s Marshall McLuhan coined the term 'global village' to capture the fact that the earth was becoming a smaller place. This trend has continued as we can now travel between major centres anywhere in the world within a day. Even more impressive is our ability to communicate instantaneously around the world through the Internet and exchange volumes of data and information on just about any subject imaginable in just a few seconds. Through telecommunications we can speak as easily to someone across the ocean as we can to our neighbour across the fence. We also have 24-hour radio and television coverage of news stories that continuously brings formerly distant events directly into our homes.

Globalisation has affected fire management in many ways of which three will be highlighted. First, wildland fire, which was once considered a local and seasonal issue, has become a global and year-round news item. It is rare to go more than a few months without watching dramatic live news footage of flames racing through forests, bushlands or grasslands, encroaching upon towns and the associated panic of citizens as they evacuate their communities or attempt in valiant but often futile ways to protect their homes. The same scenes that are played out in North America are also happening in southern Europe and Australia, constantly keeping wildfire, or at least its socio-economically destructive aspects, on our minds. Recognising their common issues and challenges, wildland fire-management organisations meet on a regular basis to share knowledge and experiences through regional networks set up by the United Nations Food and Agriculture Organization as well as its International Strategy for Disaster Reduction. There have been four world conferences on wildland fire management that have helped bring together fire researchers, policy makers and managers, the most recent being held in Spain in 2007 attended by over 1500 representatives from 88 countries. We have the Global Fire Monitoring Centre, an information and monitoring facility located in Freiburg, Germany, that connects national and international agencies involved in land-use planning, fire and disaster management as well as research and policy. There are also hundreds of other smaller and more frequently held meetings, workshops, study tours and interchanges that facilitate the worldwide exchange of information and ideas. Through the sharing of best practices and lessons learnt there is a trend towards global convergence of policies and approaches for forest fire management. One example is the widespread use of the Canadian Forest Fire Weather Index System (see Chapter 7) which has been implemented or slightly adapted by at least 20 countries worldwide. Another example is the growing number of jurisdictions around the world who are adopting the American-developed Incident Command System as the preferred mechanism for managing large wildland fires and other incidents. A third example is the coalescence of international and interdisciplinary research teams, such as occurred for the ARCTAS (Arctic Research of the Composition of the Troposphere from Aircraft and Satellites) project, to study wildland fire as part of a much larger issue.

Second, for many years during extreme fire events, suppression resources have been shared among different agencies within countries. There have also been long-standing resource-sharing agreements between neighbouring countries such as Canada and the USA as well as Australia and New Zealand. Then in 2000, during a period of extreme fires in the western USA, when all traditional sources of available firefighting resources were tapped out (including those in Canada), the USA requested and received fire-suppression resources from Australia and New Zealand. This was the first major, trans-equatorial transfer of firefighting resources and 3 years later the USA reciprocated by sending crews to Australia. Subsequently, at the International Wildland Fire Summit in Sydney in 2003, an agreement was adopted to facilitate the establishment of mutual aid agreements between countries worldwide. Although just a few are in place there are an increasing number of bilateral agreements being negotiated between countries and more can be anticipated as demand for resources increases and capacity in any one nation or organisation declines.

Third, as a result of globalisation we are seeing transnational organisations, both private sector and not-for-profit, become integral players in fire management taking over roles traditionally reserved solely for governments. One example is in Chile where private sector forest-management companies have created highly productive, fast-growing plantations and in order to protect their investment have established world-class fire information systems and maintain a professional, well-equipped fire-suppression organisation. In contrast, the public land in Chile is managed by the government and has a very limited budget and fewer resources for addressing fire prevention, suppression or use. A second example is The Nature Conservancy which in the USA is one of the largest private owners of wildland and frequently uses fire to manage the health and productivity of its lands (Pyne 2004). The implications of this shift from public management to private or non-governmental management may be viewed positively or negatively depending on one's perspective; the key point is that as it becomes harder and harder to fund fire-management activities through public expenditures (due to other priorities) there is the potential for a continuing shift away from having governments, most of which are democratically elected, actively managing land and fire.

Economics

Many economic systems, ranging from local to global, tend to be cyclical in nature; however, when sudden changes occur, especially downturns, they often seem to be unexpected by economic experts, investors and the general public. For example, in mid 2008, it would have been difficult to find large numbers of people who were concerned about the global economy. Financial markets around the world had been on a steady incline since early 2003 following the fallout from the accounting scandals in several large companies economic growth driven by massive levels of development in China and India was strong and steady; and the price of oil was at an all-time high. Six months later the status of the world's economy was very different. Financial markets saw dramatic falls, governments supplied trillions of dollars to help ailing financial institutions and industries as well as to bolster consumer confidence and spending, and oil fell to one-quarter of its peak price. Early in 2009, it was being called the most significant recession since the great depression of the 1930s. For those who believed in unlimited economic growth as the path to prosperity this was a major crisis; whereas to others who focused on long-term sustainability this was a long overdue major correction in a system that had not given due consideration to natural capital and the environment (e.g. Daly & Cobb 1994).

Regardless of one's view, during an economic downturn fire-management agencies are often subject to reductions in their fixed budgets in order to make funds available to stimulate the economy. Historically, the first casualties of budget cuts to fire-management programmes tend to be prevention, research and development, and hazard reduction activities, partly because they are less urgent and less visible than fire suppression. There also tends to be less appetite for prescribed burning programmes as risk-averse administrators claim that even one escaped fire would be too costly. On the other hand, in extreme instances governments sometimes draw upon the tactics successfully employed by US President Franklin D. Roosevelt in the 1930s who sought to kick-start the economy by funding major infrastructure projects and other initiatives. It was this type of thinking that saw the Civilian Conservation Corps build miles and miles of fuel breaks, access roads and fire towers in the USA. Conversely, once an economy has rebounded and prosperous times have returned, agencies seem to have greater success in hiring more

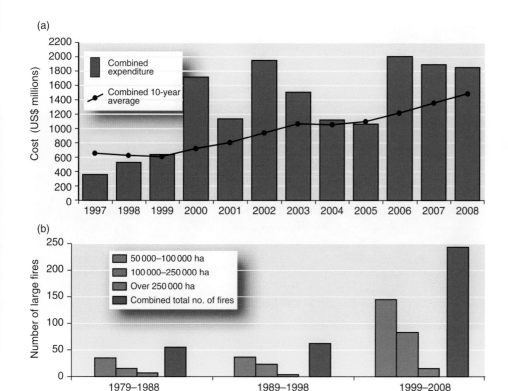

Figure 8.5

(a) Fire-suppression expenditures by federal agencies in the USA from 1997–2008 and the numbers and size of the fires involved (figures corrected for inflation). Data from the Quadrennial Fire Review (2009). (b) Number of large wildfires on federal lands in the USA from 1979–2008. Data from the National Wildfire Coordinating Group (2009).

firefighters, purchasing new equipment, upgrading ageing aircraft, undertaking new research and funding innovative on-the-ground management initiatives. The challenge for wildland fire-management agencies is to find creative ways to effectively deliver their long-term programmes given the continuing trend of short-term fluctuations in our economic environment.

Another major economic concern is that the cost of fire suppression (Fig. 8.5a) has been on the rise (National Wildfire Coordinating Group 2009). Two key causes are (1) the increase in the number of large fires (Fig. 8.5b), which account for the majority of fire-suppression expenditures, and (2) the growing number of fires in the WUI, which are often an order of magnitude more expensive than similar fires affecting only natural resource values, like timber or watersheds. It is also important to recognise that fire suppression is most effective when fires are relatively small and of low to moderate intensity (Hirsch & Martell 1996); however, when fires become large and very intense the impact of even the best and most powerful suppression resources on the spread and intensity of a wildfire is virtually unnoticeable. Sending air tankers and heavy equipment to halt the forward progress of a raging wildfire is akin to trying to stop a hurricane, tornado or tsunami – it is just not physically possible. There remains, however, tremendous political and public pressure on fire managers to do so regardless of the effectiveness of their efforts or the cost (Wuerthner 2006). Even those on the frontlines are beginning to openly discuss the futility of attempting to control large wildfires

and the extreme costs associated with needing to be seen to be doing something as noted in a 29 July 2008 *Los Angeles Times* article entitled 'Air tanker drops in wildfires often just for show'.

Some fire managers are attempting to tackle the fire-cost issue by introducing a risk management approach to decision making. Examples include:

- Using spatial and temporal data to develop probability distributions about potential fire weather, fire behaviour and ignitions, and then proactively adjust the number and position of fire-suppression resources so that they are readily available for initial attack, thereby helping to reduce the likelihood of escape and the number of large fires.

- Delaying the contracting of expensive resources like heavy helicopters (which can cost over US$ 10 000 an hour to operate) at the height of a fire's intensity and instead wait for a change in conditions (which usually means rain) and then investing in less-expensive ground crews to do fireline mop-up.

- Focusing limited resources on point protection of key values (such as communities, watersheds, critical wildlife habitat, plantations) using methods such as high-volume sprinkler systems and burnouts (see Chapter 7) rather than attempting perimeter control (Box 8.2).

Shifting the mindset away from doing 'everything humanly possible' to stop a wildfire will not be easy as it relates directly to our human psyche; but the rising cost of fire suppression combined with new fiscal realities resulting from the global recession of 2008–9 may, in fact, be the catalyst that stimulates a comprehensive rethinking of the approaches used in large fire management.

Figure 8.6
Photos during and after a wildfire at Mesa Verde National Park, Colorado, 29 July–4 August 2002. The staff were extremely proud of the fact that there was no loss of life and minimal structural damage during this fire: 'Mesa Verde has done a lot of things right . . . all the years of fuel reduction; months of preplanning for evacuation, structural fire planning, water management; with days of readiness, training, drills, and hours of urban interface tactics paid off on July 29th'. Courtesy of US National Park Service.

Box 8.2. Protecting valuables

Value protection is most effective when proactive hazard reduction measures are combined with an efficient suppression/response system. An excellent case in point is from Mesa Verde Colorado (see Fig. 8.6) where in 2003 severe wildfires occurred but losses of life and property were minimal.

Another interesting economic issue to highlight is the growing amount of litigation related to wildland fire. This is considered an economic matter because it is a legal means of determining who pays for the undesirable effects of wildland fire. Given the increasing impact that wildfire is having on personal property, businesses and human health and safety, it is unlikely that the number of court cases will decline. A rise in litigation would, in turn, make it more difficult to recruit individuals to work in senior decision-making positions (including on incident command teams) due to concerns about being sued. It would also reinforce the risk-averse nature of governments providing even more justification for not implementing active and effective prescribed burn programmes even where they are ecologically essential and cost-effective (Yoder *et al.* 2004).

A final economic driver that appears poised to hit the mainstream is the trading of carbon credits. This would involve businesses buying and selling credits for actions taken to reduce or offset greenhouse gas (GHG) emissions from fossil fuels (Sandor *et al.* 2003). This could have significant implications for forest and fire management since forests are generally carbon sinks while they are actively growing and carbon sources when they burn or decompose after insect and disease attacks (Kurz *et al.* 2008). Not surprisingly, a concept being put forward by some governments and environmental groups is that to help achieve GHG emission targets, jurisdictions should set aside large tracts of forest and protect them from fire. In some ecosystems, like rainforests, this may be an appropriate strategy; however, in fire-dependent ecosystems, especially those subject to high-intensity, stand-replacing crown fires, the feasibility of this approach is highly suspect. Interestingly there are also discussions occurring about the creation of trading systems related to biodiversity and there is rapidly growing interest in the use of forest biomass, live or dead, for bioenergy. We do not know if these or other economic influences will materialise but suffice to say that as society's values-at-risk expand and/or change there will be significant ramifications for why, where and how wildland fire is managed.

Climate change

In 1988 James Hansen, scientist and Head of NASA's Goddard Institute for Space Studies, stood with little fanfare before the US Senate Energy Committee in Washington, DC and for the first time political leaders were made aware of the issue of climate change and alerted to the serious impacts it could have on our world. Since then the topic of climate change has emerged from primarily a scientific discussion to be the pre-eminent environmental concern of our day and one that is frequently and passionately discussed by people in all walks of life. Along with melting sea ice, polar bears and shrinking glaciers, wildfire is a poster child for climate change due to the influence of extreme weather on fire activity. Over the past two decades there has been a proliferation of studies around the world on the potential effects of climate

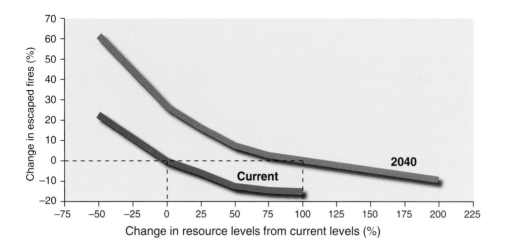

Figure 8.7
Per cent change in the number of escaped fires (from current level) with changes in resource levels for both current weather and under a climate with doubled atmospheric carbon dioxide (the year 2040). This analysis was conducted for the province of Ontario using the Level of Protection Analysis System and showed that, to maintain the current level of escape fires under one set of projected conditions, the provincial suppression resources would have to double (i.e. a 100% increase). From Wotton & Stocks (2006). Canadian Forest Service (Natural Resources Canada), reproduced with permission.

change on fire (e.g. Cary 2002, Westerling *et al.* 2006, Alcamo *et al.* 2007, Flannigan *et al.* 2008) and although high spatial variability is expected, many scenarios suggest that under a $2\times CO_2$ environment (i.e. a doubling of pre-industrial carbon dioxide levels to reach 550 ppm, which is projected to occur by about 2040) fire seasons could be longer, droughts may be lengthier and more severe, and ignitions (both human- and lightning-caused) may increase. Assuming current approaches and levels of protection, this could lead to a sharp jump in escape fires (Fig. 8.7) and area burnt and increased damage to resources, including ecosystems as they burn under conditions beyond what is considered historically 'natural'.

Climate variability and climate change are having a major impact on forest insects and diseases, which in turn could augment fuel loads and horizontal fuel continuity (Volney & Hirsch 2005). A prime example is the rapid expansion of the mountain pine beetle (*Dendroctonus ponderosae*) infested lodgepole pine stands (*Pinus contorta*) in western Canada, which in a period of 10 years grew from just a few hundred hectares to over 10 million hectares (Taylor *et al.* 2006). Similar outbreaks have occurred in the western USA and when combined with temperature and drought-induced mortality (van Mantgem *et al.* 2009) has raised major concerns for fire-management organisations and communities situated in the midst of large tracts of dead or dying trees.

Fire's role as an agent of change is likely to be accentuated under a changing climate. More high-intensity fires and more severe post-fire conditions may cause certain sites to be uninhabitable to their traditionally dominant species and more frequent fire may accelerate changes in ecosystems. For example, Flannigan *et al.* (2002) suggest that 'increased fire frequency at the grassland-aspen parkland-boreal forest transition in western Canada ... may hasten the conversion of boreal forest to aspen parkland and aspen parkland to grassland'. This has tremendous implications for forest- and fire-management policies and practices, especially in parks and protected areas. Millar *et al.* (2007) point out that under a changing climate returning ecosystems to what was considered natural may no longer be feasible and if this is true there may be major implications for ecosystem restoration programmes, especially those that use prescribed fire.

Another important aspect of the fire and climate-change relationship is the fact that forests contain a large proportion of the carbon held in all terrestrial ecosystems, and therefore wildland fires can produce a significant amount of GHGs, which in turn contribute to climate change. In Canada, Amiro *et al.* (2001) calculated that in a severe fire season the amount of carbon dioxide emissions from wildfires was equivalent to 75% of the human-produced emissions in the country. Global numbers are not available but massive fire seasons like 1983 in Borneo, 1998 in Indonesia and 2003 in Russia when over 23 million hectares burnt (an area almost the size of Britain) undoubtedly released large amounts of GHG in the atmosphere and even into the upper troposphere and lower stratosphere where their effects are further amplified (Fromm *et al.* 2004). Given that it takes several years or even decades before burnt forests once again become carbon sinks where they accumulate carbon faster than they lose it (e.g. Dore *et al.* 2008), the potential for a positive feedback effect is quite real. This means that as the climate changes there would be more fire and more GHG emissions which would further accelerate the rate of climate change (Kurz *et al.* 1995). As a result calls are being made for the suppression of all wildfires (though we know this is unattainable) as a quick fix by which to reduce GHG emissions. At the same time, others are advocating the elimination of prescribed burning even though in places like northern Australia it is being demonstrated that early-season prescribed fires in the savannah grasslands are helping to reduce overall emissions associated with more intense and severe dry-season fires. Therefore, decision makers around the world are encouraged to seek input and advice from those who study and analyse the dynamics, ecology, biology and economics of wildland fire from a systems perspective before introducing new policies related to forest and fire management under a changing climate.

Adaptation

The concept of sustainable development, the roots of which can be traced back to the late 1960s, seeks to balance the social, economic and environmental aspects of development (IUCN 2006). Formally defined by the United Nation's Bruntland Commission in 1987 as 'development which meets the needs of the present without compromising the ability of future generations to meet their own needs', sustainable development is a drastic departure from the preceding resource-management philosophies of sustained yield and multiple-use that sought to independently maximise long-term production of single or multiple resources independently. The notion of sustainable development was readily adopted by forest-management agencies and has manifested under the guise of ecosystem management and sustainable forest management (see Chapter 6). For wildland fire-management agencies this has required a shift in focus from fire control to a more holistic and challenging agenda that endeavours to balance the social, economic and ecological aspects of fire to maximise its benefits and minimise its detrimental impacts.

One of the key determinants of whether we as a species will be able to truly achieve sustainable development and flourish in the future (utopia) or will flounder and collapse (dystopia) is likely to be our ability to adapt to rapidly changing circumstances when there is little or no certainty about the type, direction or magnitude of changes that will occur. Adaptiveness (or adaptation), according to Carol Pierce Colfer (2005) in her book *The Complex Forest*, is:

> a naturally occurring process whereby people and systems evolve, changing their behaviour over time, to adjust to changing circumstances. But adaptiveness can also be a more conscious attempt to structure feedback by means of ongoing monitoring, into a given system so that changes can be made purposefully in an institutionalized manner.

Adaptation, therefore, involves social learning and is one of the primary methods by which individuals, industries, communities and governments can effectively deal with social, economic and environmental changes. The importance of adaptation has been significantly heightened by the already evident impacts of climate change (Flannery 2006). In the realm of climate change, adaptation is viewed as complementary to mitigation of GHG emissions and is defined by the Intergovernmental Panel on Climate Change (Parry *et al.* 2007) as 'adjustment in natural or human systems in response to actual or expected climatic stimuli or their effects, which moderates harm or exploits beneficial opportunities'. There are various types of adaptation, including anticipatory, autonomous and planned adaptation and it is generally acknowledged that proactive adaptation will be more effective and less costly in the long run than responding to changes when they occur (Stern 2007, Lemmen *et al.* 2008).

Adaptation generally occurs at a personal, local or community level but is enabled by perceptions, philosophies and policies from individual to global levels. A unique feature of adaptation is that in the face of what appears to be a global process well beyond a single person's control (such as climate change or the world economy), there is an opportunity to assess information, explore scenarios, develop strategies and take action to improve our individual adaptive capacity. This, in turn, can affect others and collectively enhance the resilience of larger social and ecological systems, thereby increasing the capacity to cope with surprises (Folke *et al.* 2002).

Linked to the basic principles of adaptation is the concept of adaptive management. First proposed for use in natural resource management by Holling (1978) and Walters (1986), adaptive management 'treats on-the-ground actions and policies as hypotheses from which learning derives, which, in turn, provides the basis for changes in subsequent actions and policies' (Stankey *et al.* 2005). It is an empowering approach that provides a responsible and pragmatic way forward

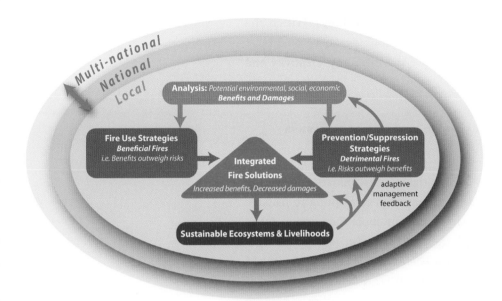

Figure 8.8
Myers (2006) shows how adaptive management is a part of his concept of Integrated Fire Management. Reproduced with permission.

when little or no certainty exists about causes or consequences. In Fig. 8.8, R.L. Myers illustrates how the concept of adaptive management could be applied in a wildland fire context (Myers 2006). His framework for 'Integrated Fire Management' emphasises the fact that fire has beneficial roles and detrimental impacts depending on the socio-economic and environmental circumstances and that many fire-management decisions are and will continue to be made with incomplete knowledge and limited experience. Through their work with the 'Resilience Network', Gunderson & Holling (2002) take the concept of adaptation to a new level by providing an integrative theory of transformation in complex systems, be they human or natural. Referred to as 'panarchy' the theory is based on an adaptive cycle that integrates exploitation, conservation, release (or creative destruction) and reorganisation. The authors indicate it is 'cross-scale, interdisciplinary, and dynamic [in] nature' and strives to 'rationalize the interplay between change and persistence, between the predictable and unpredictable'. Applicable to changes in economic, ecological and institutional systems, they say 'the purpose of theories such as panarchy is not to explain what is; it is to give a sense of what might be. We cannot predict the specifics of future possibilities, but we might be able to define the conditions that limit or expand those future possibilities.'

Innovation

Closely tied to adaptation is innovation, which by one definition is simply a new way of doing something, be it thinking, processing, producing products or running organisations. Innovation is occurring all the time and often results in incremental improvements to existing systems. Occasionally, however, there is an innovation that provides a totally new and fresh approach to a seemingly entrenched way of thinking or insurmountable issue and redefines how we do things. Within the wildland fire-management community there are numerous examples of this, of which three are identified here to illustrate we have the ingenuity, creativity and willingness to be innovative and adapt.

Banff National Park, Canada: a systems-design approach to ecosystem and fire management

Systems design is a concept that first appeared in the interdisciplinary engineering community in the mid 1970s. It is especially useful when attempting to manage systems that have a high level of complexity due to multiple dimensions, interactions and feedbacks and is consistent with the principles of ecosystem management (see Cortner & Moote 1999 or McCormick 1999 for detailed explanations) and the techniques used in landscape analysis and design (Diaz & Bell 1997). Professor Ed Jernigan, of the University of Waterloo in Canada, states that '*systems* is how we know the world; *design* is how we change the world'. The staff at Banff National Park are, likely without knowing it, pioneers in the use of a systems-design approach to wildland fire management.

Banff National Park (NP) is a UNESCO world heritage site located in the heart of the Canadian Rockies. Established in 1885 it is Canada's oldest National Park and receives almost 5 million visitors annually. Over 6600 km^2 in size, Banff NP is managed by Parks Canada, a Canadian federal government agency whose employees must balance and reconcile complex and at times conflicting social, economic and environmental issues (Fig. 8.9) under the close watch of neighbouring jurisdictions, environmental groups, the private sector and the general public.

During most of the twentieth century, Parks Canada's fire policy called for the elimination of all forest fires. Similar to the philosophies of parks and reserves in other countries the desire was to preserve nature's beauty in perpetuity just as it existed in the present moment. This meant that Parks Canada's fire-management ambitions were no different than those of other Canadian fire agencies, namely to prevent and suppress all wildfires. Then in 1968 a major wildfire occurred at Vermillion Pass in Kootenay NP, just outside of Banff NP. This was the first major wildfire in the region in over 30 years and raised some serious doubts within Parks Canada staff and management about their fire-control capacity and capability. It also coincided with the emergence of the modern environmental movement and the resurrection of previously discounted theories about the natural role of fire in ecosystems put forward by individuals such as John Muir and Aldo Leopold in the USA and Stan Rowe in Canada. Over the next decade Parks Canada's management philosophies and approaches were passionately debated, internally and in public forums, leading to new policies focused on ecological integrity (Parks Canada 1979) and an innovative and integrated fire-management policy aptly titled *Keepers of the Flame* (Parks Canada 1989).

Figure 8.9
A prescribed fire in Banff National Park burns adjacent to the trans-Canada highway and just a few kilometres from the town of Banff. Photograph by Randy Komar.

Creating a new fire-management policy is one challenge, putting it into practice is another, especially with few financial resources, limited experience in the use of fire and intense public scrutiny evaluating, and often criticising, every action. In addition to Parks Canada's new ecological integrity objectives, Banff NP staff needed to consider how to protect the town of Banff, which in the mid 1980s contained about 5000 permanent residents and would swell to a population five times that size during the peak summer tourist season. There were also business needs to take into account including key transportation routes (both the Trans-Canada Highway and the Canadian-Pacific Railway run through the park), world-renowned ski resorts and timber and other natural resource values bordering the park. Therefore, the conundrum faced in Banff NP was that, on one hand, it was essential for fire to be present on the landscape to ensure its ecological integrity but, on the other hand, it was necessary to protect key economic and social values at risk situated in relatively homogeneous and volatile fuels. This meant that neither the modus operandi of fire elimination was appropriate nor the natural fire or 'let-burn' technique that was in vogue in places such as Yellowstone and Glacier National Parks in the USA. Caught between a rock and a hard place, the creative and pragmatic staff at Banff NP were able to find a third way; that is, a systems-design approach that would enable them to manage the park's ecosystems to meet their ecological integrity objectives *and* simultaneously minimise the social and economic risks.

There were five major activities or programmes undertaken by Banff NP staff to achieve their objectives. First, they began by gathering detailed information about the historical role of fire (human- and lightning-caused) in the park and learnt about the effects of fire on various elements of the ecosystem. Second, they enhanced their initial attack programme to reduce the likelihood of unwanted wildfires threatening people and structures. Third, they instituted Canada's most active pre-scribed burning programme, sensibly recognising that this may not have been totally natural but that it was much easier to control the risks associated with planned prescribed fires than those accompanying free-burning wildfires. Fourth, through the use of science, sound analysis, persuasion and considerable persistence, Banff NP staff established the first major wildland–urban interface (WUI) fuels management programme in the country. And finally, a major communications and relations campaign was put in place to educate local citizens, business owners and tourists, as well as Parks Canada's own management and staff about the important role of fire in maintaining the health of the park's ecosystems and how it needed to be managed.

Today the perspectives and practices initiated in Banff NP are commonplace in most National Parks in Canada (Parks Canada 2005). Furthermore, due to Parks Canada's leadership, the importance of the ecological and biological role of fire in the vast majority of Canada's ecosystems is now fully accepted and this has been formalised within the Canadian Wildland Fire Strategy (CCFM 2005) which states that 'fire is an essential ecological process that contributes to the productivity, health, and biodiversity of the forests'.

State Farm Insurance – changing the odds

Turn on the television or radio in any season and there is a good chance that you will see pictures of communities somewhere in the world being threatened, people being evacuated and homes being burnt down. In the aftermath of these incidents invariably there will be interviews with devastated homeowners who indicate that they never thought it could happen to them but vow to rebuild. At this juncture it would seem logical that homeowners and government officials would consider if the home should be rebuilt in an area prone to wildland fires and/or what precau-tions should be taken to minimise the risk of it burning down again (Fig. 8.10). Furthermore, one would expect those who provide insurance for these homes would want to help prevent the same mistakes from being made twice. In other words, just like a person living in a large urban centre is told about the importance of having a smoke detector and fire extinguisher, and receives a discount on their policy because of that, a homeowner in the WUI would be advised to have a class A roof (i.e. one that is nearly impossible to ignite) or to carefully clear flammable vegetation away from the house for an adequate distance (as described in Chapter 6), and receive a reduced rate on their policy.

To those of us outside of the insurance world and with little understanding of actuarial data and how it is used to determine insurance rates, this seems to be pure common sense. In fact fire managers often raise the need for insurance companies to play a role in enticing WUI residents to reduce the vulnerability of their

Figure 8.10
Residents taking action to reduce the
vulnerability of their community to wildfire.
Courtesy of Firewise Communities® Program.

homes. For their part, however, insurance companies have found that wildland fires, although dramatic when viewed on television, account for much less than 1% of their total losses worldwide. They pay out much more for events such as floods, hailstorms and kitchen cooking fires. Strictly speaking, from a claims perspective it makes little economic sense for them to worry about wildland fire, thus it is all the more amazing that one company, State Farm Insurance, has made a concerted effort to incorporate wildland fire considerations into their homeowners' fire insurance programme.

The story of State Farm Insurance, one of the largest insurance firms in the USA, and its interest in wildland fire began in 2002 in Colorado. A regional Loss Mitigation Coordinator, reviewing property loss information from recent wildfires in his home state, was struck by the amount of physical damage that had occurred, sensed that this could be just the tip of the proverbial iceberg, and decided to look more seriously at the 'level of exposure' of State Farm's policy-holders to wildland fire. At first he began working at the local level, connecting with fire and forestry officials to learn more about the issues and what could be done about them. He then amassed relevant data, including the WUI 'red zone map' of Colorado showing its most wildfire-vulnerable areas, and was persuasively able to convince people in his corporate systems department to load it into their computer so he could overlay it with the locations of their policy-holders' properties and generate estimated 'exposure' reports. With this new-found knowledge in hand, he approached his regional executive, depicted the severity of the issue in terms they could understand, explained the relevance of it to the company and their clients, and provided a proposed solution with an associated business case. Highly impressed, the regional executive approved a new programme that would help educate State Farm agents and policy-holders about the risks from wildland fire and how to proactively reduce them. Although the executives were aware that wildland fire losses for the company as a whole were low, the data showed that within Colorado and other neighbouring states, risks were high and potential losses were not trivial. Furthermore, they recognised the bigger-picture issue, namely that this was not just about the bottom-line numbers on an accounting ledger but that they truly did have a responsibility to live up to their company's slogan – like a good neighbour, State Farm is there – and inform their policy-holders that building homes out of highly flammable materials in hazardous locations was neither in their best interest nor the company's, and that alternative solutions did exist. With this support in place, wildland fire considerations were directly incorporated into the existing re-inspection programme and within 3 years over 26 000 properties had been inspected in the region (which included Arizona, Colorado, Nevada, New Mexico, Utah and Wyoming). Policy-holders were informed of any problems with their property, possible solutions and/or directed to fire experts and then given 18 months to make the changes (after which time the company had the option to not renew the policy if changes were not made).

Propelled by the passion, ingenuity and dedication of the regional Loss Mitigation Coordinator, championed by the regional executive and supported by corporate head office, the programme quickly grew to include State Farm's operations in their other regions in the western USA (Hodges 2007). As Malcolm Gladwell points out in his bestselling book the *Tipping Point* (2000), little things – when they are the right things in the right combination – can make a big difference, and that is certainly true in the case of State Farm Insurance and the initiative of the Loss Mitigation Coordinator.

Australia – respecting wildfire and taking responsibility

On Ash Wednesday 1983 following months of drought, multiple fires pushed by extreme winds burnt thousands of homes and killed 75 people. This was the largest number of recorded deaths from a wildfire in the twentieth century and at the time these fires were ranked among the most severe in the world in the modern era of fire management.[1] Many of the victims died trying to

---

1. In February 2009 an even worse incident occurred when wildfires burning near Melbourne, Australia under 46 °C temperatures and strong, shifting winds destroyed thousands of homes and tragically caused the death of an estimated 173 people, many of whom died while trying to flee the fires. The citizens of Australia, who are more accustomed to wildfires that any other people on the planet, were severely shaken by this incident. In response the Premier of the State of Victoria established a Royal Commission to investigate the causes and responses to the bushfires including examining the government's bushfire management policies and practices. Although similar to the Ash Wednesday Fires of 1983, the wildland fire challenges in Australia appear to be increasingly daunting. However, the commitment of Australians to learn and adjust from these events is a sign of a highly adaptive society.

escape the fires which rapidly spread from the bushlands into the suburban developments. An extensive review of this incident concluded that there was a need for a major change in the attitudes and behaviour of those living in or near flammable bushlands. Given the frequency and ferocity of fire in Australia, referred to by some as the continent of fire, it was deemed essential that Australians understand and respect the potential for wildfires (or bushfires as they are referred to), that they learn to live harmoniously with fire, and accept an increased degree of responsibility for the safety of themselves, their families and the protection of their property.

This dramatic change in philosophy resulted in a widespread public education campaign aimed at changing people's perceptions and ultimately their behaviour. One result was the creation of a variety of material for the education of elementary school-aged children as well as adults. An outstanding example (Fig. 8.11) is a short children's book by Ronald Jones and Michael Salmon entitled *The Three Little Diggers: A Modern-day Scarytale (For Children to Read to Adults)*. A creative and colourful adaptation of the fairy tale of *The Three Little Pigs*, it skilfully depicts some of the common values and attitudes of urbanites who want to 'build a house in the country', cleverly characterises the nature of bushfires in Australia and how they are a natural part of the ecosystem but occasionally a threat to people (i.e. 'the big, bad bushfire ... but it wasn't always bad ... the plants and animals learnt to put up with him'); and pointedly concludes with examples of what not to do and what to do in a fire-prone wildland environment in order to live 'safely ever after'.

There was also a major change in philosophy regarding evacuations. Rather than legislate that people must evacuate their homes and leave its protection to professional firefighters as is currently still practised in many countries like Canada and the USA, the 'prepare, leave early, or stay and defend' concept (described in Chapter 7) was developed. Although not this official position policy, almost reached a mindset in some locales, and was considered relatively successful by many fire managers. Key ingredients for it to function effectively in Australia or any other country include:

- Having residents who are fully wildfire aware and respect what can happen under extreme conditions

- Investing time and resources into teaching individuals and community groups about how to protect themselves proactively as well as what to do before and during an emergency situation arises

- Ensuring the public media is actively involved before, during and after a wildland fire emergency by helping to disseminate factual information about conditions and incidents rather than sensationalising them

Although encouraging people who live in the wildland–urban interface to take more responsibility for their own safety and protection is not unreasonable by

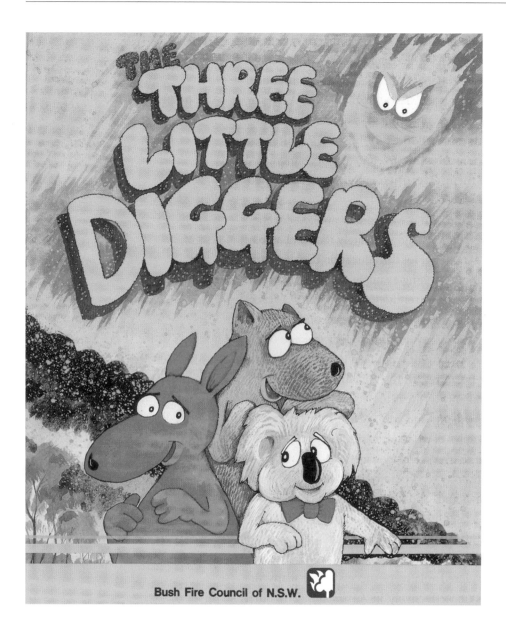

Figure 8.11
An example of the type of educational material developed after the 1983 Ash Wednesday fires in Australia to help wildland–urban interface residents, young and old, learn about bush fires and how to better protect their homes. Courtesy of the NSW Rural Fire Service.

any yardstick, the 'prepare, leave early or stay and defend' philosophy is not without its detractors and is one of the main areas of focus in the reviews of the 2009 fires.

The future – ours for the making

> Our ability to anticipate the future is very limited, while the future itself contains infinite possibilities. (Cornish 2004)

Imagine the year is 2040 and the 'World Sustainability Council', established in 2015, has just released its report for the past year. The numbers for the 'Global Sustainability Index' (GSI) are excellent and show that for the first time in recorded history the social, economic and ecological needs of present and future generations are in balance and being adequately met. Each year the report highlights an area where major progress has been made since the inception of the GSI in 2015 and this year it is focused on wildland fire and its management. The report indicates that the vast majority of the world is in a state where fire-cognisant and knowledgeable people are living in Firewise or FireSmart homes and communities (see Chapter 6) that are situated in fire-adapted ecosystems. It goes on to specify the following factors as the key ingredients that have fostered the changes that have occurred since the second decade of the twenty-first century.

First, a significant change occurred in people's view of the world and their role in it. They adopted systems-thinking and began to see themselves as part of nature, rather than apart from it, and gained an entirely new respect for wildland fire and its value in meeting ecological and human needs. People also took responsibility for increasing the resilience of ecosystems, chose to be the true keepers of the flame by increasing their wise and appropriate use of fire to meet their obligations as global stewards.

Second, major shifts occurred within both public and private institutions. Organisations radically restructured themselves so that they were able to adapt to rapidly changing circumstances and employ an adaptive approach to ecosystem and wildland fire management. They increased their flexibility and moved away from rule- and regulation-based management to approaches based on principles, collective responsibility and individual accountability. Specific actions were also taken to increase civic discourse in decision making. This involved using highly inclusive and collaborative processes that promoted citizen engagement such as deliberative democracy.

Third, innovation occurred at an amazing rate and allowed critical, emerging and unexpected challenges to be addressed in a timely manner. This was facilitated through the establishment of formal and informal networks that enabled the open exchange of best practices and lessons learnt from successful and failed experiments and trials. Innovation was further cultivated through post-normal science conducted via transdisciplinary projects combining the social sciences, the biophysical sciences and the humanities. There was also the integration of science, policy and programmes as well as initiatives, such as embedded science, that brought researchers, policy makers and practitioners together onto the same team.

Fourth, organisations realised the value, both tangible and intangible, of being proactive in their planning and invested heavily in environmental scanning, trend and scenario analysis, visioning and strategic foresight. This resulted in the incorporation of cross-scale and inter-generational factors into their decision-making process.

Last, and potentially most importantly, there was the emergence of strong leaders in many organisations and at many different levels, from local to global, all at about the same time. Information technology played an indispensible role in connecting these key influencers, allowing their creative yet disparate advances in fire management to coalesce and rapidly reach a tipping point whereby they became the norms within the wildland fire-management community and also received widespread public acceptance and support.

To some, the above idealised scenario may seem unlikely or even impossible. To some, it is difficult to see how the obstacles we face on a daily basis and the inertia in our current systems can move us past the feeling that we are simply like the mythical Greek King Sisyphus, who was condemned to ceaselessly roll a huge boulder up a steep hill simply to have it roll back down again before he reached the top. To some, the changes needed in human behaviour (including embracing uncertainty and viewing the future with curiosity; courageously taking risks and thinking outside-of-the-box to develop and implement plans based on our best but incomplete knowledge; taking actions conscious of avoiding irreversible acts, carefully monitoring the effects of our actions, then adjusting our perceptions, policies and practices as required to ensure we truly live in harmony with nature and one another) are simply a pipe-dream or the whims of wishful thinkers.

Let us ask ourselves, however, who in 1869, when Jules Verne wrote about humanity one day going to the moon, would have imagined that this was anything other than pure fiction? Likewise, in the midst of the cold war few foresaw that through a series of small, seemingly inconsequential events the dreaded Berlin wall would suddenly collapse ending decades of extreme international tension. And third, who in 1955 when Rosa Parks was required by law to give up her seat on a public bus in Mobile, Alabama because of her race (but refused) would have dreamed that within their lifetime an African-American would become the President of the United States of America? In all of these cases, there were at least a few who believed that the seemingly impossible could be accomplished, a few who believed in the potential of the human mind and spirit to overcome challenging obstacles and insurmountable odds placed on us by old societal norms and views. Today, if we step back from the barrage of headlines and soundbites provided to us by the mass media and are able to reflect on the world around us, it is clear that within the wildland fire-management community we have an abundance of dedicated, insightful and creative people who are willing, able and beginning to take the steps necessary to work collectively to prevent, suppress and actively use fire in order to achieve sustainable land and resource management objectives that are in the best interests of present and future generations. It is true that 'the future ain't what it used to be' which means that it really is ours for the making and the time to begin is now.

Further reading

A wide selection of books have been written about many aspects of fire. For good accessible accounts, the following are recommended.

Cottrell, W. H. Jr (2004) *The Book of Fire* (2nd edn). Missoula, MT: Mountain Press.

Peluso, B. A. (2007) *The Charcoal Forest: How Fire Helps Animals & Plants.* Missoula, MT: Mountain Press.

Pyne, S. J. (2008) *Year of the Fires: The Story of the Great Fires of 1910.* Missoula, MT: Mountain Press.

Other books on fire history by Stephen Pyne are also recommended and can be found in the references below.

The following may be of use for general reference:

Arno, S. F. & Allison-Bunnell, S. (2001) *Flames in our Forest: Disaster or Renewal?* Washington, DC: Island Press.

Bond, W. J. & van Wilgen, B. W. (1996) *Fire and Plants.* London: Chapman & Hall.

Booysen, P. de V. & Tainton, N. M. (1984) *Ecological Effects of Fire in South African Ecosystems.* Berlin: Springer.

Bowman, D. M. J. S. (2000) *Australian Rainforests, Islands of Green in a Land of Fire.* Cambridge: Cambridge University Press.

Chandler, C., Cheney, P., Thomas, P. & Williams, D. (1983) *Fire in Forestry. Vol 1. Forest Fire Behavior and Effects. Vol. 2. Forest Fire Management and Organization.* New York: Wiley.

De la Heras, J., Brebbia, C. A., Viegas, D. & Leone, V. (2008) *Modelling, Monitoring and Management of Forest Fires.* Southampton: WIT Press.

Gill, A. M., Groves, R. H. & Noble, I. R. (1981) *Fire and the Australian Biota.* Canberra: Australian Academy of Science.

Goldammer, J. G. (1990) *Fire in the Tropical Biota.* Berlin: Springer.

Johnson, E. A. & Miyanishi, K. (2001) *Forest Fires: Behaviour and Ecological Effects.* San Diego, CA: Academic Press.

Kozlowski, T. T. & Ahlgren, C. E. (1974) *Fire and Ecosystems.* New York, NY: Academic Press.

Kull, C. A. (2004) *Isle of Fire: The Political Ecology of Landscape Burning in Madagascar.* Chicago, IL: University of Chicago Press.

Moreno, J. M. (1998) *Large Forest Fires.* Leiden: Backhuys.

Pyne, S. J., Andrews, P. L. & Laven, R. D. (1996) *Introduction to Wildland Fire* (2nd edn). New York, NY: Wiley.

Wein, R. W. & MacLean, D. A. (1983) *The Role of Fire in Northern Circumpolar Ecosystems.* SCOPE 18. Chichester: Wiley.

Whelan, R. J. (1995) *The Ecology of Fire.* Cambridge: Cambridge University Press.

References used in the text

Ainsworth, A. & Kauffman, J. B. (2009) Response of native Hawaiian woody species to lava-ignited wildfires in tropical forests and shrublands. *Plant Ecology*, **201**: 197–209.

Alaback, P., Veblen, T. T., Whitlock, C. *et al.* (2003) Climatic and human influences on fire regimes in temperate forest ecosystems in North and South America. In *How Landscapes Change: Human Disturbance and Ecosystem Fragmentation in the Americas*, ed. Bradshaw, G. A. & Marguet, P. A. Berlin: Springer, pp. 49–87.

Alcamo, J., Moreno, J. M., Novaky, B. *et al.* (2007) Europe. In *Climate Change 2007: Impacts, Adaptation and Vulnerability*, ed. Change, M. L., Parry, O. F., Canziani, J. P. *et al.* Contribution of Working Group II to the Fourth Assessment Report of the Intergovernmental Panel on Climate. Cambridge: Cambridge University Press, pp. 541–580.

Alexander, M. E. (1982) Calculating and interpreting forest fire intensities. *Canadian Journal of Botany*, **60**: 349–357.

Alexander, M. E. & De Groot, W. J. (1988) *Fire Behavior in Jack Pine Stands as Related to the Canadian Forest Fire Weather Index System*. Poster with text. Northern Forestry Centre, Edmonton, AB: Canadian Forest Service.

Alley, R. B. (2000) The Younger Dryas cold interval as viewed from central Greenland. *Quaternary Science Reviews*, **19**: 213–226.

Alley, R. B. (2004) *GISP2 Ice Core Temperature and Accumulation Data. IGBP PAGES/World Data Center for Paleoclimatology Data Contribution Series #2004–013*. Boulder, CO: NOAA/NGDC Paleoclimatology Program.

Amacher, A. J., Barrett, R. H., Moghaddas, J. J. & Stephens, S. L. (2008) Preliminary effects of fire and mechanical fuel treatments on the abundance of small mammals in the mixed-conifer forest of the Sierra Nevada. *Forest Ecology and Management*, **255**: 3193–3202.

Amatullia, G., Peréz-Cabelloa, F. & de la Riva, J. (2007) Mapping lightning/human-caused wildfires occurrence under ignition point location uncertainty. *Ecological Modelling*, **200**: 321–333.

Amiro, B. D., Todd, J. B., Wotton, B. M. *et al.* (2001) Direct carbon emissions from Canadian forest fires, 1959–1999. *Canadian Journal of Forest Research*, **31**: 512–525.

Anderson, A. J. (1989) *Prodigious Birds: Moas and Moa-hunting in Prehistoric New Zealand*. Cambridge: Cambridge University Press.

Andrew, M. H. (1986) Use of fire for spelling monsoon tall grass pasture grazed by cattle. *Tropical Grasslands*, **20**: 69–78.

Andrews, P. L. (1986) *BEHAVE: Fire Behaviour Prediction and Fuel Modelling System. BURN Subsystem Part 1*. General Technical Report INT-194. Intermountain Forest and Range Experiment Station, Ogden, UT: United States Department of Agriculture Forest Service.

Andrews, P. L. & Rothermel R. C. (1982) *Charts for Interpreting Wildland Fire Behavior Characteristics*. General Technical Report INT-131. Intermountain Forest and Range Experiment Station, Ogden, UT: United States Department of Agriculture Forest Service.

Anon (1970) *Forest Fire Fighting Fundamentals*. USDA and California Division of Forestry.

Anon (1973) *Forest Enemies*. Ottawa, ON: Canadian Forestry Service, Department of the Environment.

Anon (2001) *Global Forest Fire Assessment 1990–2000*. Working Paper 55. Rome: Forestry Department, Food and Agriculture Organization of the United Nations (available from www.fao.org/documents).

Arno, S. F. & Sneck, K. M. (1977) *A Method for determining Fire History in Coniferous Forests of the Mountain West*. General Technical Report INT-42. Intermountain Forest and Range Experiment Station, Ogden, UT: United States Department of Agriculture Forest Service.

Ashton, D. H. (1986) Viability of seeds of *Eucalyptus obliqua* and *Leptospermum juniperinum* from capsules subjected to a crown fire. *Australian Forestry*, **49**: 28–35.

Auld, T. D. & Denham, A. J. (2006) How much seed remains in the soil after a fire? *Plant Ecology*, **187**: 15–24.

Axelson, J. N., Alfaro, R. I. & Hawkes, B. C. (2009) Influence of fire and mountain pine beetle on the dynamics of lodgepole pine stands in British Columbia, Canada. *Forest Ecology and Management*, **257**: 1874–1882.

Babrauskas, V. (2002) Ignition of wood: a review of the state of the art. *Journal of Fire Protection Engineering*, **12**: 163–189.

Baskin, Y. (1999) Yellowstone fires: a decade later. *Bioscience*, **2**: 93–97.

Belcher, C. M., Collinson, M. E. & Scott, A. C. (2005) Constraints on the thermal energy released from the Chicxulub impactor: new evidence from multi-method charcoal analysis. *Journal of the Geological Society, London*, **162**: 591–602.

Berg, A., Ehnstrom, B., Gustafsson, L. *et al.* (1994) Threatened plant, animal, and fungus species in Swedish forests: distribution and habitat associations. *Conservation Biology*, **8**: 718–731.

Bergeron, Y., Gauthier, S., Kafka, V., Lefort, P. & Lesieur, D. (2001) Natural fire frequency for the eastern Canadian boreal forest: consequences for sustainable forestry. *Canadian Journal of Forest Research*, **31**: 384–391.

Berner, R. A. (1999) Atmospheric oxygen over Phanerozoic time. *Proceedings of the National Academy of Sciences USA*, **96**: 10955–10957.

Billings, R. F., Clarke, S. R., Mendoza, V. E. *et al.* (2004) Bark beetle outbreaks and fire: a devastating combination for Central America's pine forests. *Unasylva*, **217**: 15–21.

Binkley, D. (1993) Group report: impacts of fire on ecosystems. In *Fire in the Environment: The Ecological, Atmospheric and Climatic Importance of Vegetation Fires*, ed. Crutzen, P. J. & Goldammer, J. G. New York, NY: John Wiley, pp. 359–372.

Bond, W. J., Honig, M. & Maze, K. E. (1999) Seed size and seedling emergence: an allometric relationship and some ecological implications. *Oecologia*, **120**: 120–132.

Bond, W. J. & Midgley, J. J. (1995) Kill thy neighbour: an individualistic argument for the evolution of flammability. *Oikos*, **73**: 79–85.

Bond, W. J., Woodward, F. I. & Midgley, G. F. (2004) The global distribution of ecosystems in a world without fire. *New Phytologist*, **165**: 525–538.

Booysen, P. de V. & Tainton, N. M. (1984) *Ecological Effects of Fire in South African Ecosystems*. Berlin: Springer.

Borchert, M., Johnson, M., Schreiner, D. S. & Vander Wall, S. B. (2003) Early postfire seed dispersal, seedling establishment and seedling mortality of *Pinus coulteri* (D. Don) in central coastal California, USA. *Plant Ecology*, **168**: 207–220.

Born, W. & Stocks, B. J. (2006) Canadian fire management infrastructure. In *Canadian Wildland Fire Management Strategy: Background Syntheses, Analyses, and Perspectives*, ed. Hirsch, K. G. & Fuglem, P. (technical coordinators). Northern Forestry Centre, Edmonton, AB: Canadian Forest Service, Canadian Council of Forest Ministers, Natural Resources Canada, pp. 57–71.

Bova, A. S. & Dickinson, M. B. (2005) Linking surface-fire behavior, stem heating, and tissue necrosis. *Canadian Journal of Forest Research*, **35**: 814–822.

Bowman, D. (2005) Understanding a flammable planet – climate, fire and global vegetation patterns. *New Phytologist*, **165**: 341–345.

Bradshaw, R. H. W. & Zackrisson, O. (1990) A two thousand year history of a northern Swedish boreal forest stand. *Journal of Vegetation Science*, **1**: 519–528.

Bradstock, R. A. (2008) Effects of large fires on biodiversity in south-eastern Australia: disaster or template for diversity? *International Journal of Wildland Fire*, **17**: 809–822.

Bradstock, R. A., Williams, J. E. & Gill, M. A. (2002) *Flammable Australia: The Fire Regimes and Biodiversity of a Continent*. Cambridge: Cambridge University Press.

Brewer, J. S., Cunningham, A. L., Moore, T. P., Brooks, R. M. & Waldrup, J. L. (2009) A six-year study of fire-related flowering cues and coexistence of two perennial grasses in a wet longleaf pine (*Pinus palustris*) savanna. *Plant Ecology*, **200**: 141–154.

Brotak, E. A. & Reifsnyder, W. E. (1977) An investigation of the synoptic situations associated with major wildland fires. *Journal of Applied Meteorology*, **16**: 867–870.

Brown, A. A. & Davis, K. P. (1973) *Forest Fire: Control and Use* (2nd edn). New York, NY: McGraw-Hill.

Brown, J. K. & DeByle, N. V. (1987) Fire damage, mortality, and suckering in aspen. *Canadian Journal of Forest Research*, **17**: 1100–1109.

Buddle, C. M., Langor, D. W., Pohl, G. R. & Spence, J. R. (2006) Arthropod responses to harvesting and wildfire: implications for emulation of natural disturbance in forest management. *Biological Conservation*, **128**: 346–357.

Buddle, C. M., Spence, J. R. & Langor, D. W. (2000) Succession of boreal forest spider assemblages following wildfire and harvesting. *Ecography*, **23**: 424–436.

Buhk, C. & Hensen, I. (2008) Seed longevity of eight species common during early postfire regeneration in south-eastern Spain: a 3-year burial experiment. *Plant Species Biology*, **23**: 18–24.

Burton, T. A. (2005) Fish and stream habitat risks from uncharacteristic wildfire: observations from 17 years of fire-related disturbances on the Boise National Forest, Idaho. *Forest Ecology and Management*, **211** (Special Issue 1–2): 140–149.

Butler, K. (2008) Interpreting charcoal in New Zealand's palaeoenvironment – what do those charcoal fragments really tell us? *Quaternary International*, **184**: 122–128.

Byram, G. M. (1959) Combustion of forest fuels. In *Forest Fire: Control and Use*, ed. Davis, K. P. New York, NY: McGraw Hill, pp. 61–123.

Calef, M. P., McGuire, A. D. & Chapin III, F. S. (2008) Human influences on wildfire in Alaska from 1988 through 2005: an analysis of the spatial patterns of human impacts. *Earth Interactions*, **12**: 1–17.

Campbell, I. D. & Campbell, C. (2000) Late Holocene vegetation and fire history at the southern boreal forest margin in Alberta,

Canada. *Palaeogeography, Palaeoclimatology, Palaeoecology*, **164**: 263–280.

Carcaillet, C., Bergman, I., Delorme, S., Hornberg, G. & Zackrisson, O. (2007) Long-term fire frequency not linked to prehistoric occupations in northern Swedish boreal forest. *Ecology*, **88**: 465–477.

Carter, M. C. & Foster, C. D. (2004) Prescribed burning and productivity in southern pine forests: a review. *Forest Ecology and Management*, **191**: 93–109.

Cary, G. L. (2002) The importance of a changing climate for fire regimes in Australia. In *Flammable Australia*, ed. Bradstock, R. A., Williams, J. E. & Gill, A. M. Cambridge: Cambridge University Press, pp. 26–46.

CCFM (2005) *Canadian Wildland Fire Strategy: A Vision for an Innovative and Integrated Approach to Managing the Risks*. Ottawa, ON: Canadian Council of Forest Ministers.

Certini, G. (2005) Effects of fire on properties of forest soils: a review. *Oecologia*, **143**: 1–10.

Chambers, F. M., Lageard, J. G. A., Boswijk, G. *et al.* (1997) Dating prehistoric bog-fires in northern England to calendar years by long-distance cross-matching of pine chronologies. *Journal of Quaternary Science*, **12**: 253–256.

Chandler, C., Cheney, P., Thomas, P. Trabaud, L. & Williams, D. (1983). *Fire in Forestry. Vol. 1. Forest Fire Behavior and Effects*. New York, NY: Wiley.

Chapman, H. H. (1952) The place of fire in the ecology of pines. *Bartonia*, **26**: 39–44.

Cheney, N. P. (1981) Fire behaviour. In *Fire and the Australian Biota*, ed. Gill, A. M., Groves, R. H. & Noble, I. R. Canberra: Australian Academy of Science, pp. 151–175.

Choung, Y., Lee, B.-C., Cho, J.-H. *et al.* (2004) Forest response to the large-scale east coast fires in Korea. *Ecological Research*, **19**: 43–54.

Clark, J. S. (1988) Effects of climate change on fire regime in northeastern Minnesota. *Nature*, **334**: 233–235.

Clarke, P. J. & Dorji, K. (2008) Are trade-offs in plant resprouting manifested in community seed banks? *Ecology*, **89**: 1850–1858.

Climent, J., Tapias, R., Pardos, J. A. & Gill, L. (2004) Fire adaptations in the Canary Islands pine (*Pinus canariensis*). *Plant Ecology*, **171**: 185–196.

Cochrane, M. A. & Barber, C. P. (2009) Climate change, human land use and future fires in the Amazon. *Global Change Biology*, **15**: 601–612.

Colfer, C. J. P. (2005) *The Complex Forest: Communities, Uncertainty, and Adaptive Collaborative Management*. Washington, DC: RFF Press.

Collins, S. L., Glenn, S. M. & Gibson, D. J. (1995) Experimental analysis of intermediate disturbance and initial floristic composition: decoupling cause and effect. *Ecology*, **76**: 486–492.

Conroy, R. J. (1996) To burn or not to burn? A description of the history, nature and management of bushfires within Ku-ring-gai Chase National Park. *Proceedings of the Linnean Society of New South Wales*, **116**: 79–95.

Cope, M. J. & Chaloner, W. G. (1980) Fossil charcoal as evidence of past atmospheric composition. *Nature*, **283**: 647–649.

Cornish, E. (2004) *Futuring: The Exploration of the Future*. Bethesda, MD: World Future Society.

Cortner, H. J. & Moote, M. A. (1999) *The Politics of Ecosystem Management*. Washington, DC: Island Press.

Cottrell, W. H. Jr (2004) *The Book of Fire* (2nd edn). Missoula, MT: Mountain Press.

Coulson, B., Curry, G., Tchakerian, M., Gan, J. & Smith, C. T. (2005) *Utilization of plant biomass generated from Southern pine beetle outbreaks*. The 2005 ESA Annual Meeting and Exhibition, 15–18 December 2005, Fort Lauderdale, FL.

Cowling, R. M., Byron, B. L. & Pierce, S. M. (1987) Seed bank dynamics of four co-occurring *Banksia* species. *Journal of Ecology*, **75**: 289–302.

Daly, H. & Cobb, J., Jr (1994) *For the Common Good: Redirecting the Economy toward Community, the Environment and a Sustainable Future* (2nd edn). Boston, MA: Beacon Press.

Dennis, R., Meijaard, E., Applegate, G., Nasi, R. & Moore, P. (2001) *Impact of Human-caused Fires on Biodiversity and Ecosystem Functioning, and their Causes in Tropical, Temperate and Boreal Forest Biomes*. CBD Technical Series No. 5. Convention on Biological Diversity, Montreal, QC.

Diamond, J. (1995) Easter's end. *Discover*, **16**: 62–69.

Diamond, J. (2005) *Collapse*. New York, NY: Viking.

Díaz-Delgado, R., Lloret, F., Pons, X. & Terradas, J. (2002) Satellite evidence of decreasing resilience in Mediterranean plant communities after recurrent wildfires. *Ecology*, **83**: 2293–2303.

Diaz, N. & Bell, S. (1997) Landscape analysis and design. In *Creating a Forestry for the 21 Century: The Science of Ecosystem Management*, ed. Kohm, K. & Franklin, J. F. Washington, DC: Island Press, pp. 255–269.

Dieterich, J. H. & Swetnam, T. W. (1984) Dendrochronology of a fire-scarred ponderosa pine. *Forest Science*, **30**: 238–247.

Dixon, K. W., Roche, S. & Pate, J. S. (1995) The promotive effect of smoke derived from burnt native vegetation on seed germination of Western Australian plants. *Oecologia*, **101**: 185–192.

Dore, S., Klob, T. E., Montes-Helu, M. *et al.* (2008) Long-term impact of a stand-replacing fire on ecosystem CO_2 exchange of a ponderosa pine forest. *Global Change Biology*, **14**: 1801–1820.

Dudley, R. (1998) Atmospheric oxygen, giant paleozoic insects and the evolution of aerial locomotor performance. *Journal of Experimental Biology*, **201**: 1043–1050.

Dyer, G. (2008) *Climate Wars*. Toronto, ON: Random House.

Edwards, W. & Whelan, R. (1995) The size, distribution and germination requirements of the soil-stored seed bank of *Grevillea barklyana* (Proteaceae). *Australian Journal of Ecology*, **20**: 548–555.

FAO (2007) *Fire Management – Global Assessment 2006*. FAO Forestry Paper 151. Rome: Food and Agriculture Organization of the United Nations.

Fisher, J. T. & Wilkinson, L. (2005) The response of mammals to forest fire and timber harvest in the North American boreal forest. *Mammal Review*, **35**: 51–81.

Flannery, T. F. (2006) *The Weather Makers: The History and Future Impact of Climate Change*. Toronto, ON: HarperCollins.

Flannigan, M. D. & Harrington, J. B. (1988) A study of the relation of meteorological variables to monthly provincial area burned by wildfire in Canada (1953–80). *Journal of Applied Meteorology*, **27**: 441–452.

Flannigan, M. D., Krawchuk, M. A., de Groot, W. J., Wotton, B. M. & Gowman, L. M. (2009) Implications of changing climate for global wildland fire. *International Journal of Wildland Fire*, **18**: 483–507.

Flannigan, M. D., Stocks, B. J., Turetsky, M. R. & Wotton, B. M. (2008) Impact of climate change on fire activity and fire management in the circumboreal forest. *Global Change Biology*, **14**: 1–12.

Flannigan, M. D., Stocks, B. J. & Weber, M. G. (2002) Fire regimes and climate change in Canadian forests. In *Fire and Climate Change in Temperate Ecosystems of the Western Americas*, ed. Veblen, T., Baker, W., Montenegro, G. & Swetnam, T. Berlin: Springer, pp. 97–119.

Flannigan, M. D. & Wotton, B. M. (1991) Lightning-induced fires in northwestern Ontario. *Canadian Journal of Forest Research*, **21**: 277–287.

Flematti, G. R., Ghisalberti, E. L., Dixon, K. W. & Trengove, R. D. (2004) A compound from smoke that promotes seed germination. *Science*, **305**(5686): 977.

Flinn, M. A. & Wein, R. W. (1977) Depth of underground plant organs and theoretical survival during fire. *Canadian Journal of Botany*, **55**: 2550–2554.

Folke, C., Carpenter, S., Elmqvist, T. *et al.* (2002) Resilience and sustainable development: building adaptive capacity in a world of transformations. *Ambio*, **31**: 437–440.

Fontaine, J. B., Donato, D. C., Robinson, W. D., Law, B. E. & Kauffman, J. B. (2009) Bird communities following high-severity fire: response to single and repeat fires in a mixed-evergreen forest, Oregon, USA. *Forest Ecology and Management*, **257**: 1496–1504.

Fortin, M.-J., Crete, M., Huot, J., Drolet, B. & Doucet, G. J. (1995) Chronoséquence après feu de la diversité de mammifères et d'oiseaux au nord de la forêt boréale québécoise. *Canadian Journal of Forest Research*, **25**: 1509–1518.

Fraver, S. (1992) The insulating value of serotinous cones in protecting pitch pine (*Pinus rigida*) seeds from high temperatures. *Journal of the Pennsylvania Academy of Science*, **65**: 112–116.

Fromm, M., Bevilacqua, R., Stocks, B. & Servranckx, R. (2004) New directions: eruptive transport to the stratosphere: add fire-convection to volcanoes. *Atmosphere Environment*, **38**: 163–165.

Gashaw, M. & Michelsen, A. (2002) Influence of heat shock on seed germination of plants from regularly burnt savanna woodlands and grasslands in Ethiopia. *Plant Ecology*, **159**: 1–83.

Giglio, L., van der Werf, G. R., Randerson, J. T., Collatz, G. J. & Kasibhatla, P. (2006) Global estimation of burned area using MODIS active fire observations. *Atmospheric Chemistry and Physics*, **6**: 957–974.

Gill, A. M. (1981) Post-settlement fire history in Victorian landscapes. In *Fire and the Australian Biota*, ed. Gill, A. M., Groves, R. H. & Noble, I. R. Canberra: Australian Academy of Science, pp. 77–98.

Gill, A. M. & Allan, G. (2008) Large fires, fire effects and the fire-regime concept. *International Journal of Wildland Fire*, **17**: 688–695.

Gill, A. M., Christian, K. R. & Moore, P. H. R. (1987) Bushfire incidence, fire hazard and fuel reduction burning. *Australian Journal of Ecology*, **12**: 299–306.

Gill, A. M. & Moore, P. H. R. (1996) *Ignitibility of Leaves of Australian Plants*. Contract Report to the Australian Flora Foundation. Canberra: Centre for Plant Biodiversity Research, CSIRO Plant Industry.

Gill, A. M. & Moore, P. H. R. (1998) Big versus small fires: the bushfires of Greater Sydney, January 1994. In *Large Forest Fires*, ed. Moreno, J. M. Leiden: Backhuys, pp. 49–68.

Gladwell, M. (2000) *The Tipping Point: How Little Things can Make a Big Difference*. Boston, MA: Little and Brown Company.

Glasspool, I. J., Edwards, D. & Axe, L. (2004) Charcoal in the Silurian as evidence for the earliest wildfire. *Geology*, **32**: 381–383.

Goh, K.-T., Schwela, D., Goldammer, J. G. & Simpson, O. (1999) *Health Guidelines for Vegetation Fire Events: Background Papers*. Geneva: World Health Organization (available from www.who.int/peh/air/vegetationfirbackgr.htm).

Goldammer, J. G. (1993) *Feuer und Waldentwicklung in den Tropen und Subtropen*. Berlin: Birkhäuser.

González-Pérez, J. A., González-Vila, F. J., Almendros, G. & Knicker, H. (2004) The effect of fire on soil organic matter – a review. *Environment International*, **30**: 855–870.

Gordon, J. A. (2002) *Comprehensive Emergency Management for Local Governments: Demystifying Emergency Planning*. CT: Rothstein Associates, Brookfield.

Gott, B. (2005) Aboriginal fire management in south-eastern Australia: aims and frequency. *Journal of Biogeography*, **32**: 1203–1208.

Goubitz, S., Werger, M. J. A. & Ne'eman, G. (2003) Germination response to fir-related factors of seeds from non-serotinous and serotinous cones. *Plant Ecology*, **169**: 195–204.

Granstrom, A. (2001) Fire management for biodiversity in the European boreal forest. *Scandinavian Journal of Forest Research Supplement*, **3**: 62–69.

Green, K. & Sanecki, G. (2006) Immediate and short-term responses of bird and mammal assemblages to a subalpine wildfire in the Snowy Mountains, Australia. *Austral Ecology*, **31**: 673–681.

Greenberg, C. H. & Waldrop, T. A. (2008) Short-term response of reptiles and amphibians to prescribed fire and mechanical fuel reduction in a southern Appalachian upland hardwood forest. *Forest Ecology and Management*, **255**: 2883–2893.

Grier, C. E. (1975) Wildfire effects on nutrient distribution and leaching in a coniferous ecosystem. *Canadian Journal of Forest Research*, **5**: 599–607.

Groen, A. H. & Woods, S. W. (2008) Effectiveness of aerial seeding and straw mulch for reducing post-wildfire erosion, north-western Montana, USA. *International Journal of Wildland Fire*, **17**: 559–571.

Grove, S. J. (2001) Extent and composition of dead wood in Australian lowland tropical rain forests with different management histories. *Forest Ecology and Management*, **154**: 35–53.

Gunderson, L. H. & Holling, C. S. (2002) *Panarchy: Understanding Transformations in Human and Natural Systems.* Washington, DC: Island Press.

Haeussler, S. & Bergeron, Y. (2004) Range of variability in boreal aspen plant communities after wildfire and clear-cutting. *Canadian Journal of Forest Research*, **34**: 274–288.

Haines, D. A. (1982) Horizontal roll vortices and crown fires. *Journal of Applied Meteorology*, **21**: 751–763.

Hall, B. L. (2007) Precipitation associated with lightning-ignited wildfires in Arizona and New Mexico. *International Journal of Wildland Fire*, **16**: 242–254.

Hammond, P. M. (1994) Practical approaches to the estimation of the extent of biodiversity in speciose groups. *Philosophical Transactions of the Royal Society, London*, **B345**: 119–136.

Haney, A., Apfelbaum, S. & Burris, J. M. (2008) Thirty years of post-fire succession in a southern boreal forest bird community. *American Midland Naturalist*, **159**: 421–433.

Hanley, M., Unna, J. & Darvill, B. (2003) Seed size and germination response: a relationship for fire-following plant species exposed to thermal shock. *Oecologia*, **134**: 18–22.

Harmon, M. E., Franklin, J. F. & Swanson, F. J. (1986) Ecology of coarse woody debris in temperate ecosystems. *Advances in Ecological Research*, **15**: 133–302.

Heliövarra, K. & Väisänen, R. (1984) Effects of modern forestry on northwestern European forest invertebrates: a synthesis. *Acta Forestalia Fennica*, **189**: 1–32.

Hély, C., Bergeron, Y. & Flannigan, M. D. (2000) Effects of stand composition on fire hazard in mixed-wood Canadian boreal forest. *Journal of Vegetation Science*, **11**: 813–824.

Hernández, D. L. & Hobbie, S. E. (2008) Effects of fire frequency on oak litter decomposition and nitrogen dynamics. *Oecologia*, **158**: 535–543.

Hille, M. & den Ouden, J. (2005) Charcoal and activated carbon as adsorbate of phytotoxic compounds – a comparative study. *Oikos*, **108**: 202–207.

Hirsch, K. G., Corey, P. N. & Martel, D. L. (1998) Using expert judgment to model initial attack crew effectiveness. *Forest Science*, **44**: 539–549.

Hirsch, K. G. & Martell, D. L. (1996) A review of initial attack fire crew productivity and effectiveness. *International Journal of Wildland Fire*, **6**: 199–215.

Hobson, K. A. & Schieck, J. (1999) Changes in bird communities in boreal mixedwood forest: harvest and wildfire effects over 30 years. *Ecological Applications*, **9**: 849–863.

Hodges, V. (2007) Homeowner education program aims to reduce wildfire risk. *Disaster Safety Review*, 2007: 10–11, 15.

Holdaway, R. N. & Jacomb, C. (2000) Rapid extinction of the moas (Aves: Dinornithiformes): model, test, and implications. *Science*, **287**: 2250–2254.

Holling, C. S. (1978) *Adaptive Environmental Assessment and Management.* London: John Wiley.

Homer-Dixon, T. (2001) *The Ingenuity Gap.* Toronto, ON: Vintage Canada.

Hungerford, R. D., Frandsen, W. H. & Ryan, K. C. (1995). Ignition and burning characteristics of organic soils. *Proceedings of the 19th Tall Timbers Fire Ecology Conference: Fire in Wetlands: A Management Perspective, Tallahassee, Florida*, **19**: 78–91.

Hunter, M. L. (1990) *Wildlife, Forests and Forestry: Principles of Managing Forests for Biological Diversity.* Upper Saddle River, NJ: Prentice-Hall.

Iniguez, J. M., Swetnam, T. W. & Yool, S. R. (2008) Topography affected landscape fire history patterns in southern Arizona, USA. *Forest Ecology and Management*, **256**: 295–303.

IPCC (2007) *Climate Change 2007: The Physical Science Basis.* Contribution of Working Group I to the Fourth Assessment Report of the Intergovernmental Panel on Climate Change. Cambridge: Cambridge University Press.

IUCN (2006) *The Future of Sustainability: Re-thinking Environment and Development in the Twenty-first Century.* Report of the IUCN Renowned Thinkers Meeting, 29–31 January 2006. Gland: International Union for the Conservation of Nature.

Jacobs, J. M., Spence, J. R. & Langor, D. W. (2007) Influence of forest succession and dead wood qualities on boreal saproxylic beetles. *Agriculture and Forest Entomology*, **9**: 3–16.

Johnson, B. (1984) *The Great Fire of Borneo: Report of a Visit to Kalimantan-Timur a Year Later, May 1984.* Godalming: World Wildlife Fund.

Johnson, E. A. (1992) *Fire and Vegetation Dynamics: Studies from the North American Boreal Forest.* Cambridge: Cambridge University Press.

Johnson, E. A. & Gutsell, S. L. (1994) Fire frequency models, methods and interpretations. *Advances in Ecological Research*, **25**: 239–287.

Johnson, E. A., Miyanishi, K. & Bridge S. R. J. (2001) Wildfire regime in the boreal forest and the idea of suppression and fuel buildup. *Conservation Biology*, **15**: 1554–1557.

Johnson, P. S. (1975) Growth and structural development of red oak sprout clumps. *Forest Science*, **21**: 413–418.

Jones, R. (1975) The Neolithic Palaeolithic and hunting gardeners: man and land in the Antipodes. In *Quaternary Studies*, ed. Suggate, P. & Cresswell, M. M. Wellington: Royal Society of New Zealand, pp. 21–34.

Jones, R. & Salmon, M. (1985) *The Three Little Diggers.* Sydney: Bush Fire Council of New South Wales.

Kafka, V., Gauthier, S. & Bergeron, Y. (2001) Fire impacts in the boreal forest: study of a large wildfire in western Quebec. *International Journal of Wildland Fire*, **10**: 119–127.

Karlsson, M., Caesar, S., Ahnesjö, J. & Forsman, A. (2008) Dynamics of colour polymorphism in a changing environment: fire melanism and then what? *Oecologia*, **154**: 715–724.

Karthikeyan, S., Balasubramanian, R. & Iouri, K. (2006) Particulate air pollution from bushfires: human exposure and possible health effects. *Journal of Toxicology and Environmental Health, Part A, Current Issues*, **69**: 1895–1908.

Kaval, P., Loomis, J. & Seidl, A. (2007) Willingness-to-pay for prescribed fire in the Colorado (USA) wildland urban interface. *Forest Policy and Economics*, **9**: 928–937.

Keeley, J. E. (1998) Postfire ecosystem recovery and management: the October 1993 large fire episode in California. In *Large Forest Fires*, ed. Moreno, J. M. Leiden: Backhuys, pp. 69–90.

Keeley, J. E. & Fotheringham, C. J. (2000) Role of fire in regeneration from seed. In *Seeds: The Ecology of Regeneration in Plant Communities*, ed. Fenner, M. Oxford: Commonwealth Agricultural Bureau International, pp. 311–330.

Kershaw, A. P., Clark, J. S., Gill, A. M. & D'Costa, D. M. (2002) A history of fire in Australia. In *Flammable Australia*, ed. Bradstock, R. A., Williams, J. E. & Gill, M. A. Cambridge: Cambridge University Press, pp. 1–25.

Kiss, L. & Magnin, F. (2003) The impact of fire on some Mediterranean land snail communities and patterns of post-fire recolonization. *Journal of Molluscan Studies*, **69**: 43–53.

Korb, J. E., Johnson, N. C. & Covington, W. W. (2004) Slash pile burning effects on soil biotic and chemical properties and plant establishment: recommendations for amelioration. *Restoration Ecology*, **12**: 52–62.

Kramer, K., Groena, T. A. & van Wieren, S. E. (2003) The interacting effects of ungulates and fire on forest dynamics: an analysis using the model FORSPACE. *Forest Ecology and Management*, **181**: 205–222.

Kurz, W. A., Apps, M. J., Stocks, B. J. & Volney, W. J. A. (1995) Global climate change: disturbance regimes and biospheric feedbacks of temperate and boreal forests. In *Biotic Feedbacks in the Global Climate System: Will the Warming Speed the Warming?* ed. Woodwell, G. M. & Mackenzie, F. Oxford: Oxford University Press, pp. 119–133.

Kurz, W. A., Stinson, G., Rampley, G. J., Dymond, C. & Neilson, E. T. (2008) Risk of natural disturbances makes future contribution of Canada's forests to the global carbon cycle highly uncertain. *Proceedings of the National Academy of Sciences USA*, **105**: 1551–1555.

Kuwana, K., Sekimoto, K., Saito, K. *et al.* (2007) Can we predict the occurrence of extreme fire whirls? *AIAA Journal*, **45**: 16–19.

Lamont, B. B., Le Maitre, D. C., Cowling, R. M. & Enright, N. J. (1991) Canopy seed storage in woody plants. *Botanical Review*, **57**: 277–317.

Langner, A. & Siegert, F. (2009) Spatiotemporal fire occurrence in Borneo over a period of 10 years. *Global Change Biology*, **15**: 48–62.

Larjavaara, M., Pennanen, J. & Tuomi, T. J. (2007) Lightning that ignites forest fires in Finland. *Agricultural and Forest Meteorology*, **132**: 171–180.

Lawson, B. D. & Armitage, O. B. (2008) *Weather Guide to the Canadian Forest Fire Danger Rating System.* Northern Forestry Centre, Edmonton, AB: Canadian Forest Service, Natural Resources Canada.

LeDuc, S. D. & Rothstein, D. E. (2007) Initial recovery of soil carbon and nitrogen pools and dynamics following disturbance in jack pine forests: a comparison of wildfire and clearcut harvesting. *Soil Biology and Biochemistry*, **39**: 2865–2876.

Lee, P. C., Crites, S., Nietfeld, M., Van Nguyen, H. & Stelfox, J. B. (1997) Characteristics and origins of deadwood material in aspen-dominated boreal forests. *Ecological Applications*, **7**: 691–701.

Lemmen, D. S., Warren, F. J., Lacroix, J. & Bush, E. (2008) *From Impacts to Adaptation: Canada in a Changing Climate 2007.* Ottawa, ON: Government of Canada.

Li, J., Loneragan, W. A., Duggin, J. A. & Grant, C. D. (2004) Issues affecting the measurement of disturbance response patterns in herbaceous vegetation – a test of the intermediate disturbance hypothesis. *Plant Ecology*, **172**: 11–26.

Lindenmayer, D. B., Burton, P. J. & Franklin, J. F. (2008a) *Salvage Logging and its Ecological Consequences.* Washington, DC: Island Press.

Lindenmayer, D. B., Wood, J. T., MacGregor, C. *et al.* (2008b) How predictable are reptile responses to wildfire? *Oikos,* **117**: 1086–1097.

Ling, P. & Storer, B. (1990) Smoke gets in your eyes. *History Today,* **40**: 6–9.

Lloret, F., Pausas, J. G. & Vilà, M. (2003) Responses of Mediterranean plant species to different fire frequencies in Garraf Natural Park (Catalonia, Spain): field observations and modelling predictions. *Plant Ecology,* **167**: 223–235.

Long, R. A., Rachlow, J. L., Kie, J. G. & Vavra, M. (2008) Fuels reduction in a western coniferous forest: effects on quantity and quality of forage for elk. *Rangeland Ecology and Management,* **61**: 302–313.

Lorimer, C. G. & Gough, W. R. (1988) Frequency of drought and severe fire weather in north-eastern Wisconsin. *Journal of Environmental Management,* **26**: 203–219.

Lui, J., Dietz, T., Carpenter, S. R. *et al.* (2007) Complexity of coupled human and natural systems. *Science,* **317**: 1513–1516.

Lunney, D., Gresser, S. M., Mahon, P. S. & Matthews, A. (2004) Post-fire survival and reproduction of rehabilitated and unburnt koalas. *Biological Conservation,* **120**: 567–575.

Lyon, J. P. & O'Connor, J. P. (2008) Smoke on the water: can riverine fish populations recover following a catastrophic fire-related sediment slug? *Austral Ecology,* **33**: 794–806.

Lyons, W. A., Nelson, T. E., Williams, E. R., Cramer, J. A. & Turner, T. R. (1998) Enhanced positive cloud-to-ground lightning in thunderstorms ingesting smoke from fires. *Science,* **282**(5386): 77–80.

Mabuhay, J., Nakagoshi, N. & Horikoshi, T. (2003) Microbial biomass and abundance after forest fire in pine forests in Japan. *Ecological Research,* **18**: 431–441.

Malakoff, D. (2002) Arizona ecologist puts stamp on forest restoration debate. *Science,* **297**(No. 5590): 2194–2196.

Marcot, B. G., Mellen, K., Livinston, S. A. & Ogden, C. (2002) *The DecAid Advisory Model: Wildlife Component.* General Technical Report PSW-GTR-181. Pacific Southwest Research Station, Albany, CA: United States Department of Agriculture Forest Service.

Margules, C. R. & Pressey, R. L. (2000) Systematic conservation planning. *Nature,* **405**: 243–253.

Martell, D. L. (2002) Wildfire regime in the boreal forest. *Conservation Biology,* **16**: 1177.

Masaka, K., Ohno, Y. & Yamada, K. (2000) Fire tolerance and the fire-related sprouting characteristics of two cool-temperate broad-leaved tree species. *Annals of Botany,* **85**: 137–142.

Masterson, G. P. R., Maritz, B. & Alexander, G. J. (2008) Effect of fire history and vegetation structure on herpetofauna in a South African grassland. *Applied Herpetology,* **5**: 129–143.

McAllum, M. J. C. (2008) *Designing Better Futures: Rethinking Strategy for a Sustainable World.* Bribie Island, Queensland: GFN Press.

McAlpine, R. S. & Hirsch, K. G. (1998) An overview of LEOPARDS: the level of protection analysis system. *Forestry Chronicle,* **75**: 615–621.

McAlpine, R. S. & Wotton, B. M. (1993) The use of fractal dimension to improve wildland fire perimeter predictions. *Canadian Journal of Forest Research,* **23**: 1073–1077.

McCaw, W. L., Smith, R. H. & Neal, J. E. (1997) Prescribed burning of thinning slash in regrowth stands of karri (*Eucalyptus diversicolor*). 1. Fire characteristics, fuel consumption and tree damage. *International Journal of Wildland Fire,* **7**: 29–40.

McCay, T. S. (2000) Use of woody debris by cotton mice (*Peromyscus gossypinus*) in a southeastern pine forest. *Journal of Mammalogy,* **81**: 527–535.

McCormick, F. J. (1999) Principles of ecosystem management and sustainable development. In *Ecosystem Management for Sustainability: Principles and Practices Illustrated by a Regional Biosphere Reserve Cooperative,* ed. Peine, J. D. New York, NY: Lewis Publishers, pp. 3–22.

McGee, T. K. (2007) Urban residents' approval of management measures to mitigate wildland–urban interface fire risks in Edmonton, Canada. *Landscape and Urban Planning,* **82**: 247–256.

McInnis, D. (1997) Burning season. *Earth,* August 1997: 36–41.

McLaren, A. C. (1959) Propagation of flames in Eucalyptus oil vapour-air mixtures. *Australian Journal of Applied Science,* **2**: 321–328.

MacLean, N. (1992) *Young Men and Fire.* Chicago, IL: University of Chicago Press.

McLennan, J. & Birch, A. (2005) A potential crisis in wildfire emergency response capability? Australia's volunteer firefighters. *Global Environmental Change Part B: Environmental Hazards,* **6**: 101–107.

MCPFE (2002) *Improvement of the Pan-European Indicators for Sustainable Forest Management.* Fourth Ministerial Conference on the Protection of Forest in Europe. Italy, Cosenza.

McRae, D. J. & Flannigan, M. D. (1990) Development of large vortices on prescribed fires. *Canadian Journal of Forest Research,* **20**: 1878–1887.

Mellars, P. A. (1976) Fire ecology, animal population and man: a study of some ecological relationships in prehistory. *Proceedings of the Prehistorical Society,* **42**: 15–46.

Mickan, F. (2006) *What Happens When Hay Heats.* Department of Primary Industries, State of Victoria, Australia, AG0206.

Midgley, J. (2002) What are the relative costs, limits and correlates of increased degree of serotiny? *Austral Ecology,* **25**: 65–68.

Millar, C. I., Stephenson, N. L. & Stephens, S. L. (2007) Climate change and forests of the future: managing in the face of uncertainty. *Ecological Applications,* **17**: 2145–2151.

Mills, G. (2005) On the sub-synoptic scale meteorology of two extreme fire weather days during the eastern Australian fires of January 2003. *Australian Meteorological Magazine,* **54**: 265–290.

Milton, S. J. & Hall, A. V. (1981) Reproductive biology of Australian acacias in the south-western Cape Province, South Africa. *Transactions of the Royal Society of South Africa*, **44**: 465–485.

Minnich, R. A. (1983) Fire mosaics in southern California and northern Baja California. *Science*, **219**(No. 4590): 1287–1294.

Minnich, R. A. (1998) Landscapes, land-use and fire policy: where do large fires come from? In *Large Forest Fires*, ed. Moreno, J. M. Leiden: Backhuys, pp. 133–158.

Miyanishi, K., Bridge, S. R. J. & Johnson, E. A. (2002) Wildfire regime in the boreal forest – reply. *Conservation Biology*, **16**: 1177–1178.

Moreira, F., Duarte, L., Catry, F. & Acacio, V. (2007) Cork extraction as a key factor determining post-fire cork oak survival in a mountain region of southern Portugal. *Forest Ecology and Management*, **253**: 30–37.

Moretti, M., Duelli, P. & Obrist, M. K. (2006) Biodiversity and resilience of arthropod communities after fire disturbance in temperate forests. *Oecologia*, **149**: 312–327.

Mouillot, F. & Field, C. B. (2005) Fire history and the global carbon budget: a $1° \times 1°$ fire history reconstruction for the 20th century. *Global Change Biology*, **11**: 398–420.

Moya, D., Saracino, A., Salvatore, R. *et al.* (2008) Anatomical basis and insulation of serotinous cones in *Pinus halepensis* Mill. *Trees*, **22**: 511–519.

Myers, R. L. (2006) *Living with Fire: Sustaining Ecosystems and Livelihoods Through Integrated Fire Management*. Tallahassee, FL: The Nature Conservancy.

Natcher, D. C., Calef, M., Huntington, O. *et al.* (2007) Factors contributing to the cultural and spatial variability of landscape burning by native peoples of interior Alaska. *Ecology and Society*, **12**: Article No. 7.

National Wildfire Coordinating Group (2009) *Quadrennial Fire Review*. Boise, ID: National Wildfire Coordinating Group Executive Board.

Neary, D. G., Ryan, K. C. & DeBano, L. F. (2005) *Fire Effects on Soil and Water*. General Technical Report RM-42. Rocky Mountain Research Station, Fort Collins, CO: United States Department of Agriculture Forest Service.

Ne'eman, G., Henig-Severa, N. & Eshel, A. (1999) Regulation of the germination of *Rhus coriaria*, a post-fire pioneer, by heat, ash, pH, water potential and ethylene. *Physiologia Plantarum*, **106**: 47–52.

Ne'eman, G., Ne'eman, R., Keith, D. A. & Whelan, R. J. (2009) Does post-fire plant regeneration mode affect the germination response to fire-related cues? *Oecologia*, **159**: 483–492.

Nichols, D. J. & Johnson, K. R. (2008) *Plants and the K-T Boundary*. Cambridge: Cambridge University Press.

Noss, R. F., Franklin, J. F., Baker, W. L., Schoennagel, T. & Moyle, P. B. (2006) Managing fire-prone forests in the western United States. *Frontiers in Ecology and the Environment*, **4**: 481–487.

Núñez, M. R. & Calvo, L. (2000) Effect of high temperatures on seed germination of *Pinus sylvestris* and *Pinus halepensis*. *Forest Ecology and Management*, **131**: 183–190.

Page, S. E., Siegert, F., Rieley, J. O. *et al.* (2002) The amount of carbon released from peat and forest fires in Indonesia during 1997. *Nature*, **420**: 61–65.

Parks Canada (1979) *Parks Canada Policy*. Ottawa, ON: Department of Environment, Government of Canada.

Parks Canada (1989) *Keepers of the Flame: Implementing Fire Management in the Canadian Parks Service*. Internal Report. Ottawa, ON: Parks Canada.

Parks Canada (2005) *National Fire Management Strategy – Parks Canada Agency*. Ottawa, ON: National Fire Management Committee, Parks Canada.

Parry, M. L., Canziani, O. F., Palutikof J. P., van der Linden, P. J. & Hanson, C. E. (2007) *Climate Change 2007: Impacts, Adaptation and Vulnerability*. Contribution of Working Group II to the Fourth Assessment Report of the Intergovernmental Panel on Climate Change. Cambridge: Cambridge University Press.

Parshall, T., Foster, D. R., Faison, E., MacDonald, D. & Hansen, B. C. S. (2003) Long-term history of vegetation and fire in pitch pine-oak forests on Cape Cod, Massachusetts. *Ecology*, **84**: 736–748.

Patterson, W. A., III, Edwards, K. J. & Maguire, D. J. (1987) Microscopic charcoal as a fossil indicator of fire. *Quaternary Science Reviews*, **6**: 3–23.

Pausas, J. G., Ouadah, N., Ferran, A., Gimeno, T. & Vallejo, R. (2003) Fire severity and seedling establishment in *Pinus halepensis* woodlands, eastern Iberian Peninsula. *Plant Ecology*, **169**: 205–213.

Penman, T. D., Binns, D. L., Shiels, R. J., Allen, R. M. & Kavanagh, R. P. (2008) Changes in understorey plant species richness following logging and prescribed burning in shrubby dry sclerophyll forests of south-eastern Australia. *Austral Ecology*, **33**: 197–210.

Penman, T. D. & Towerton, A. L. (2008) Soil temperatures during autumn prescribed burning: implications for the germination of fire responsive species? *International Journal of Wildland Fire*, **17**: 572–578.

Perrin, G. S. (1890) *Report upon the State Forests of Victoria*. Melbourne: Government Printer.

Pinchot, G. (1899) The relation of forests and forest fires. *National Geographic*, **10**: 393–403.

Pinchot, G. (1910) *The Fight for Conservation*. New York, NY: Doubleday, Page & Company.

Prideaux, G. J., Long, J. A., Ayliffe, L. K. *et al.* (2007) An arid-adapted middle Pleistocene vertebrate fauna from south-central Australia. *Nature*, **445**: 422–425.

Putz, F. E. (2003) Are rednecks the unsung heroes of ecosystem management? *Wild Earth*, **13**, 10–14.

Pyne, S. J. (1982) *Fire in America: A Cultural History of Wildland and Rural Fire*. Princeton, NJ: Princeton University Press.

Pyne, S. J. (1991) *Burning Bush. A Fire History of Australia*. New York, NY: Henry Holt.

Pyne, S. J. (1993) Keeper of the flame: a survey of anthropogenic fire. In *Fire in the Environment: The Ecological, Atmospheric, and Climatic Importance of Vegetation Fires*, ed. Crutzen, P. J. & Goldammer, J. G. Chichester: John Wiley, pp. 245–266.

Pyne, S. J. (1995) *World Fire. The Culture of Fire on Earth*. New York, NY: Henry Holt.

Pyne, S. J. (1997) *Vestal Fire. An Environmental History, Told through Fire, of Europe and Europe's Encounter with the World*. Washington, DC: University of Washington Press.

Pyne, S. J. (2001a) *Fire: A Brief History*. Washington, DC: University of Washington Press; London: British Museum.

Pyne, S. J. (2001b) *Year of the Fires: the Story of the Great Fires of 1910*. New York, NY: Penguin Books.

Pyne, S. J. (2004) *Tending Fire: Coping with America's Wildland Fires*. Washington, DC: Island Press.

Pyne, S. J. (2007) *Awful Splendour: A Fire History of Canada*. Vancouver, BC: University of British Columbia Press.

Radeloff, V. C., Hammer, R. B., Stewart, S. I. *et al.* (2005) The wildland–urban interface in the United States. *Ecological Applications*, **15**: 799–805.

Radojevic, M. (2003) Chemistry of forest fires and regional haze with emphasis on southeast Asia. *Pure and Applied Geophysics*, **160**: 157–187.

Ramsay, G. C., McArthur, N. A. & Dowling, V. P. (1996) Building in a fire-prone environment: research on building survival in two major bushfires. *Proceedings of the Linnean Society of New South Wales*, **116**: 133–140.

Reyes, O. & Trabaud, L. (2009) Germination behaviour of 14 Mediterranean species in relation to fire factors: smoke and heat. *Plant Ecology*, **202**: 113–121.

Rhemtulla, J. M., Hall, R. J., Higgs, E. S. & Macdonald, S. E. (2002) Eighty years of change: vegetation in the montane ecoregion of Jasper National Park, Alberta, Canada. *Canadian Journal of Forest Research*, **32**: 2010–2021.

Richardson, D. M. (1998) *Ecology and Biogeography of* Pinus. Cambridge: Cambridge University Press.

Rothermel, R. C. (1993) *Mann Gulch Fire: A Race that Couldn't be Won*. General Technical Report INT-299. *Intermountain Forest and Range Experiment Station, Ogden, UT: United States Department of Agriculture Forest Service*.

Ruokolainen, L. & Salo, K. (2006) The succession of boreal forest vegetation during ten years after slash-burning in Koli National Park, eastern Finland. *Annales Botanici Fennici*, **43**: 363–378.

Russell, E. W. B. (1983) Indian-set fires in the forests of the northeastern United States. *Ecology*, **64**: 78–88.

Ryan, K. C. (2002) Dynamic interactions between forest structure and fire behaviour in boreal ecosystems. *Silva Fennica*, **36**: 13–39.

Ryan, K. C. & Frandsen, W. H. (1991) Basal injury from smoldering fires in mature *Pinus ponderosa* Laws. *International Journal of Wildland Fire*, **1**: 107–118.

Sandor, R. L., Bettleheim, E. C. & Swingland, E. R. (2003) An overview of a free market approach to climate change and conservation. In *Capturing Carbon and Conserving Biodiversity: The Market Approach*, ed. Swingland, R. London: Earthscan, pp. 56–69.

Sanford, R. L., Saldarriaga, J., Clark, K. E., Uhl, C. & Herrera, R. (1985) Amazon rain-forest fires. *Science*, **227**(No. 4682): 53–55.

Sastry, N. (2002) Forest fires, air pollution, and mortality in southeast Asia. *Demography*, **39**: 1–23.

Schieck, J. & Song, S. J. (2006) Changes in bird communities throughout succession following fire and harvest in boreal forests of western North America: literature review and meta-analyses. *Canadian Journal of Forest Research*, **36**: 1299–1318.

Schmitz, A., Sehrbrock, A. & Schmitz, H. (2007) The analysis of the mechanosensory origin of the infrared sensilla in *Melanophila acuminata* (Coeloptera; Buprestidae) adduces new insight into the transduction mechanism. *Arthropod Structure and Development*, **36**: 291–303.

Schnider, S. H. (2001) What is 'dangerous' climate change? *Nature*, **411**: 17–19.

Schoennagel, T., Smithwick, E. A. H. & Turner, M. G. (2008) Landscape heterogeneity following large fires: insights from Yellowstone National Park, USA. *International Journal of Wildland Fire*, **17**: 742–753.

Schwilk, D. W., Keeley, J. E. & Bond, W. J. (1997) The intermediate disturbance hypothesis does not explain fire and diversity pattern in fynbos. *Plant Ecology*, **132**: 77–84.

See, S. W., Balasubramanian, R., Rianawati, E., Karthikeyan, S. & Streets, D. G. (2007) Characterization and source apportionment of particulate matter $\leq$ 2.5 µm in Sumatra, Indonesia, during a recent peat fire episode. *Environmental Science and Technology*, **41**: 3488–3494.

Siitonen, J. (2001) Forest management, coarse woody debris and saproxylic organisms: Fennoscandian boreal forests as an example. *Ecological Bulletins*, **49**: 11–41.

Sippola, A.-L., Siitonen, J. & Punttila, P. (2001) Beetle diversity in timberline forests: a comparison between old-growth and regeneration areas in Finnish Lapland. *Annales Zoologici Fennici*, **39**: 69–86.

Smith, N. R., Kishchuk, B. E. & Mohn, W. W. (2008) Effects of wildfire and harvest disturbances on forest soil bacterial communities. *Applied and Environmental Microbiology*, **74**: 216–224.

Stankey, G. H., Clark, R. N. & Bormann, B. T. (2005) *Adaptive Management of Natural Resources: Theory, Concepts, and Management Institutions*. General Technical Report PNW-GTR-654. Pacific Northwest Research Station, Portland, OR: United States Department of Agriculture Forest Service.

Stern, N. H. (2007) *The Economics of Climate Change: The Stern Review*. Cambridge: Cambridge University Press.

Stocks, B. J., Mason, J. A., Todd, J. B. *et al.* (2003) Large forest fires in Canada, 1959–1997. *Journal of Geophysical Research*, **107(D24)**, 8149, doi:10.1029/2001JD000484 (printed **108(D1)**, 2003).

Stocks, B. J. & Simard, A. J. (1993) Forest fire management in Canada. *Journal of Disaster Management*, **5**: 21–27.

Stokland, J. N. (2001) The coarse woody debris profile: an archive of recent forest history and an important biodiversity indicator. *Ecological Bulletins*, **49**: 71–83.

Stuart-Smith, A. K., Hayes, J. P. & Schieck, J. (2006) The influence of wildfire, logging and residual tree density on bird communities in the northern Rocky Mountains. *Forest Ecology and Management*, **231**: 1–17.

Swetnam, T. W. (1993) Fire history and climate change in giant sequoia groves. *Science*, **262**(No. 5135): 885–889.

Taylor, A. R. (1971) Lightning – agent of change in forest ecosystems. *Journal of Forestry*, **69**: 477–480.

Taylor, S. L. & MacLean, D. A. (2007) Dead wood dynamics in declining balsam fir and spruce stands in New Brunswick, Canada. *Canadian Journal of Forest Research*, **37**: 750–762.

Taylor, S. W., Carroll, A. L., Alfaro, R. I. & Safranyik, L. (2006) Forest, climate and mountain pine beetle outbreak dynamics in western Canada. In *The Mountain Pine Beetle: A Synthesis of Biology, Management, and Impacts on Lodgepine Pine*, ed. Safranyik, L. & Wilson, W. R. Pacific Forestry Centre, Victoria, BC: Canadian Forest Service, Natural Resources Canada, pp. 67–94.

Thiffault, E., Hannam, K. D., Quideau, S. A. *et al.* (2008) Chemical composition of forest floor and consequences for nutrient availability after wildfire and harvesting in the boreal forest. *Plant and Soil*, **308**: 37–53.

Thomas, P. A. & Packham, J. R. (2007) *Ecology of Woodlands and Forests*. Cambridge: Cambridge University Press.

Thomas, P. A. & Wein, R. W. (1985) Delayed emergence of four conifer species on postfire seedbeds. *Canadian Journal of Forest Research*, **15**: 727–729.

Thomas, P. A. & Wein, R. W. (1990) Jack pine establishment on ash from wood and organic soil. *Canadian Journal of Forest Research*, **20**: 1926–1932.

Tomback, D. F., Anderies, A. J., Carsey, K. S., Powell, M. L. & Mellmann-Brown, S. (2001) Delayed seed germination in whitebark pine and regeneration patterns following the Yellowstone fires. *Ecology*, **82**: 2587–2600.

Uhl, D. & Montenari, M. (2010) Charcoal as evidence of palaeo-wildfires in the Late Triassic of SW Germany. *Geological Journal*, **45**: 1–8.

Vallejo, R. (2005) Restoring Mediterranean forests. In *Forest Restoration in Landscapes: Beyond Planting Trees*, ed. Mansurain, S., Vallarui, D. & Dudley, N. New York, NY: Springer, pp. 313–319.

van Mantgem, P. J., Stephenson, N. L., Byrne, J. C. *et al.* (2009) Widespread increase of tree mortality rates in the western United States. *Science*, **323**(No. 5913): 521–524.

Van Staden, J., Brown, N. A., Jäger, A. K. & Johnson, T. A. (2000) Smoke as a germination cue. *Plant Species Biology*, **15**: 167–178.

Van Wagner, C. E. (1968) *Fire Behaviour Mechanisms in a Red Pine Plantation: Field and Laboratory Evidence*. Departmental Publication 1229. Department of Forestry and Rural Development, Forestry Branch.

Van Wagner, C. E. (1977) Conditions for the start and spread of crown fires. *Canadian Journal of Forest Research*, **7**: 23–24.

Van Wagner, C. E., Finney, M. A. & Heathcott, M. (2006) Historical fire cycles in the Canadian Rocky Mountain parks. *Forest Science*, **52**: 704–717.

Vera, F. W. M. (2000) *Grazing Ecology and Forest History*. Wallingford: CABI Publishing.

Viegas, D. X. (1998) Weather, fuel status and fire occurrence: predicting large fires. In *Large Forest Fires*, ed. Moreno, J. M. Leiden: Backhuys, pp. 31–48.

Vines, R. G. (1968) Heat transfer through bark, and the resistance of trees to fire. *Australian Journal of Botany*, **16**, 499–514.

Vining, J. & Merrick, M. S. (2008) The influence of proximity to a national forest on emotions and fire-management decisions. *Environmental Management*, **41**: 155–167.

Volney, W. J. A. & Hirsch, K. G. (2005) Disturbing forest disturbances. *Forestry Chronicle*, **81**: 662–668.

Walters, C. J. (1986) *Adaptive Management of Renewable Resources*. New York, NY: Macmillan.

Wardle, D. A., Zackrisson, O. & Nilsson, M.-C. (1998) The charcoal effect in boreal forests: mechanisms and ecological consequences. *Oecologia*, **115**: 419–426.

Webb, L. J. (1968) Environmental relationships of the structural types of Australian rain forest vegetation. *Ecology*, **49**: 296–311.

Welch, H. & Haddow, G. (1993) *The World Checklist of Conifers*. Landsman's Bookshop for the World Conifer Data Pool.

Wells, H. G. (1902) The discovery of the future. *Nature*, **65**: 326–331.

Westerling, A. L., Hidalgo, H. G., Cayan, T. & Swetnam, W. (2006) Warming and earlier spring increase western U.S. forest wildfire activity. *Science*, **313**(No. 5789): 940–943.

Wheatley, M. J. (1999) *Leadership and the New Science: Discovering Order in a Chaotic World* (2nd edn). San Francisco, CA: Berrett-Koehler Publishers.

Whelan, R. J. (1995) *The Ecology of Fire.* Cambridge: Cambridge University Press.

Whittaker, R. J., Willis, K. J. & Field, R. (2001) Scale and species richness: towards a general, hierarchical theory of species diversity. *Journal of Biogeography*, **28**: 453–470.

Wierzchowski, J., Heathcott, M. & Flannigan, M. D. (2002) Lightning and lightning fire, central cordillera, Canada. *International Journal of Wildland Fire*, **11**: 41–51.

Williams, M. (2003) *Deforesting the Earth.* Chicago, IL: Chicago University Press.

Williams, M. S. & Gove, J. H. (2003) Perpendicular distance sampling: an alternative method of sampling downed coarse woody debris. *Canadian Journal of Forest Research*, **33**: 1564–1579.

Wilson, B. G. & Witkowski, E. T. F. (2003) Seed banks, bark thickness and change in age and size structure (1978–1999) of the African savanna tree, *Burkea africana. Plant Ecology*, **167**: 151–162.

Wilson, C. C. (1977) Fatal and near fatal forest fires – the common denominators. *International Fire Chief*, **43**: 9–10, 12–15.

Wilson, C. C. & Sorenson, J. C. (1978) *Some Common Denominators of Fire Behavior on Tragedy and Near-miss Forest Fires.* Publication NA-GR-8. Broomall, PA: Northeast State and Private Forestry, United States Department of Agriculture Forest Service.

Wotton, B. M. & Stocks, B. J. (2006) Fire management in Canada: vulnerability and risk trends. In *Canadian Wildland Fire Management Strategy: Background Syntheses, Analyses, and Perspectives*, ed. Hirsch, K. G. & Fuglem, P. Northern Forestry Centre, Edmonton, AB: Canadian Forest Service, Canadian Council of Forest Ministers, Natural Resources Canada, pp. 49–55.

Wright, R. (2004) *A Short History of Progress.* Toronto, ON: House of Anansi Press.

Wuerthner, G. (2006) The flawed economics of fire suppression. In *The Wildfire Reader*, ed. Wuerthner, G. Washington, DC: Island Press, pp. 221–223.

Yoder, J., Engle, D. & Fuhlendorf, S. (2004) Liability, incentives, and prescribed fire for ecosystem management. *Frontiers in Ecology and the Environment*, **2**: 361–366.

Zanon, V., Viveiros, F., Silva, C., Hipólito, A. R. & Ferreira, T. (2008) Impact of lightning on organic matter-rich soils: influence of soil grain size and organic matter content on underground fires. *Natural Hazards*, **45**: 19–31.

Zhang, Y., He, H. S. & Yang, J. (2008) The wildland–urban interface dynamics in the southeastern U.S. from 1990 to 2000. *Landscape and Urban Planning*, **85**: 155–162.

Zielonka, T. & Niklasson, M. (2001) Dynamics of dead wood and regeneration pattern in natural spruce forest in the Tatra Mountains, Poland. *Ecological Bulletins*, **49**: 159–163.

Zwolak, R. & Foresman, K. R. (2007) Effects of a stand-replacing fire on small-mammal communities in montane forest. *Canadian Journal of Zoology*, **85**: 815–822.

Index